MATHEMATICS
FOR
PHYSICS
WITH CALCULUS

BIMAN DAS
STATE UNIVERSITY OF NEW YORK — POTSDAM

PEARSON

Prentice Hall

Upper Saddle River, NJ 07458

Library of Congress Cataloging-in-Publication Data
Das, Biman
  Mathematics for physics with calculus / Biman Das.
      p. cm.
  Includes bibliographical references.
  ISBN 0-13-191336-0 (pbk.)
  1. Mathematical physics. 2. Calculus. I. Title.

  QC23.2.D37 2005
  530.15--dc22
                                        2004002509

Associate Editor: *Christian Botting*
Senior Editor: *Erik Fahlgren*
Editor in Chief, Science: *John Challice*
Production Editor: *Kim Dellas*
Executive Managing Editor: *Kathleen Schiaparelli*
Assistant Managing Editor: *Beth Sweeten*
Manufacturing Buyer: *Alan Fischer*
Manufacturing Manager: *Trudy Pisciotti*
Cover Designer: *Bruce Kenselaar*
Editorial Assistants: *Nancy Bauer, Andrew Sobel*

© 2005 Pearson Education, Inc.
Pearson Prentice Hall
Pearson Education, Inc.
Upper Saddle River, NJ 07458

Pearson Prentice Hall ® is a trademark of Pearson Education, Inc.
Printed in Canada

 3 4 5 HI 12 11 10

ISBN 0-13-191336-0

Pearson Education LTD., *London*
Pearson Education Australia PTY, Limited, *Sydney*
Pearson Education Singapore, Pte. Ltd
Pearson Education North Asia Ltd, *Hong Kong*
Pearson Education Canada, Ltd., *Toronto*
Pearson Educacion de Mexico, S.A. de C.V.
Pearson Education—Japan, *Tokyo*
Pearson Education Malaysia, Pte. Ltd

Dedicated to the memory of my father Manindranath Das

Contents

Chapter 4
Geometry: Shapes and Plots 81
Angles, Triangles, and Quadrilaterals 81

Shapes of Objects 90

Sections marked with an asterik * indicates topics typically covered in intermediate or advanced physics.

Chapter 6
Vectors in Physics 145

Chapter 7
Differential Calculus 174

Chapter 8
Integral Calculus and Differential Equations 205

Chapter 9
Vector Calculus 239

Chapter 10
How to Be Successful in a Physics Laboratory: Data Analysis, Curve Fitting, Probability, and Statistics 261

Preface

I have written this book primarily for students who plan to take or who are presently taking calculus-based physics courses. In teaching physics courses for last several years I found that many students, who had not recently taken a course that required mathematical skill, often found themselves unable to keep pace in the physics courses because they were not familiar with the necessary mathematical tools. I have written this book with these students in mind. The book will develop necessary mathematical skills and provide students with the competence to use precalculus, calculus, vector algebra, and vector calculus, and statistical analysis of experimental data. Students taking intermediate physics, engineering, and other science courses will also find the book useful and will be able to use the book as a mathematical resource for these intermediate-level courses. Both the publisher and I realize the necessity of the book for students.

The book consists of ten chapters. Chapter 1 introduces students to physics and mathematics and discusses the role of mathematics in physics. It also discusses how to deal with the math anxiety that many students have, and how to develop good study habits, and it describes a problem-solving strategy. Chapter 2 describes scientific notation, units, prefixes, and dimensions of physical quantities. Chapters 3 through 6 describe precalculus including algebra, geometry, trigonometry, and vectors with worked out examples and exercises. Students will find how this mathematics is directly applied to physics in these chapters. Chapter 7 describes differential calculus and Chapter 8 describes integral calculus and differential equations. These two chapters show how calculus and differential equations are integral parts of physics through real-world examples and exercises from physics. Chapter 9 describes vector calculus and their applications in physics. Chapter 10 discusses statistics and elaborates on the techniques of analyzing experimental data; it also provides tips for how to be successful in a physics laboratory. Twelve appendices provide important references and topics often useful in introductory or intermediate physics.

The book emphasizes primarily the use of mathematical techniques and mathematical concepts in physics and does not go into their rigorous developments. Wherever possible, I have included examples and problems from physics so that students see how the concept of mathematics is directly applied to physics. I have written the book in such a way that students can study it to develop mathematical skills and can practice solving physics problems. There are worked-out examples and exercises in almost every section. The book can be used as a text for a one- to three-credit course (dependent on the depth of the coverage), a remedial mathematics course, or a physics recitation course for students taking a calculus-based physics course.

I would like to thank Dr. F. Eugene Dunnam (University of Florida, Gainesville), Dr. William Karstens (Saint Michael's College), Dr. Michael Broyles (Collin County Community College), Dr. David Reid (East Michigan University), Dr. Ronald J. Bieniek (University of Missouri–Rolla), Dr. Anthony

Pitucco (Pima Community College), Dr. Jenny Quan (Pasadena City College) and Dr. Dennis Rioux (University of Wisconsin–Oshkosh) for their reviews and suggestions. Special thanks are due to Dr. Michael B. Ottinger (Missouri Western State College) for his detailed comments and suggestions for improvements. I thank Mr. Christian Botting, Associate Editor, Physics and Astronomy, Prentice Hall, for his constant support and cooperation throughout the entire process. I also thank Kim Dellas, Production Editor, Prentice Hall, for her cooperation. I thank my wife Indrani Das and my daughters Debapria Das and Deea Das for their cooperation and support.

In spite of my efforts to make this book error free, it is likely that the book contains errors. I would appreciate it if I am informed about any such errors. I would also be delighted to hear any constructive comments about the book for its further improvements.

Biman Das

Note to Students: Why You Need This Book

The primary purpose of this book is to develop mathematical skills in precalculus, calculus, vectors, and statistical analysis of data that will allow you to concentrate on the physics in calculus-based physics courses. I am sure that for many of you physics courses are different from most other courses. Here, the emphasis is on understanding the physical concepts, deriving the physical laws or principles using mathematics, and applying this knowledge in solving a wide range of problems. There may be several obstacles to understanding physics and mastering problem solving. However, the primary factor that leads to your success in physics and problem solving is your competence to use mathematics with ease and accuracy. I am confident that the book will make you stronger in mathematics, including calculus, so that your learning experience in physics will be beautiful and rewarding. Note that the beauty of physics cannot be enjoyed if you are distracted and inhibited by a less-than-adequate mastery of basic mathematical skills.

You should also find the book useful in intermediate physics, engineering, and other science courses that involve problem solving, and you will be able to use the book as a mathematics resource book. Note that the book is not meant to be a substitute for standard mathematics courses that may be prerequisites for a physics course. The mathematics discussed is comparable to what you will find in a typical mathematics text. The emphasis is on enhancing those mathematical skills frequently used in physics courses. One objective of the book is to show you how the concepts of mathematics and calculus are directly applied and related to physics and real-world problems, which likely you will not find in a standard mathematics text.

To draw your attention to important concepts and formulas, I have used boxes throughout the book. I have also used the boxes to inform you of students' common mistakes or misconceptions. For your convenience, I have included important conversion factors, a list of fundamental constants, useful data, and other information in the twelve appendices. To separate the concepts that are not commonly used in introductory physics, I have used * or ** so that you will not have any problem if you want to use the book to sharpen your mathematical skills only for introductory-level physics courses. I have included most of the topics that are important for intermediate level physics in appendices.

Read the chapters carefully for mathematical concepts; study worked-out examples, and practice problems from the exercises. Learn also how to develop good study habits and how to deal with math anxiety. Good luck!

Biman Das

Chapter 1
Fun with Physics and Mathematics

The fundamental principle of natural philosophy (physics) is to attempt to reduce the apparent complex physical phenomena to some simple fundamental ideas and relations. —Einstein and Infeld

In science there is only physics, all the rest is stamp collecting. — Ernest Rutherford

1.1   How Physics Fits into Science

The word *science* was taken from the Latin word *scientia*, meaning "to know." Scientists currently define science as the systematic attainment of *knowledge* derived from observation, study, and experimentation in order to understand nature and its principles. Aristotle (384–322 B.C.) was one of the first persons to study science in a systematic manner. The greatest advancement of science came about at the beginning of the sixteenth century when people, such as Galileo (1564–1642), started looking for the relationships between physical variables and tested their ideas through experimentation.

As science grew, it was divided into two major groups: the biological sciences and the physical sciences. Biological sciences represent an organized study of living matter, while physical sciences represent an organized study of inanimate matter and energy. The physical sciences include physics, geology, chemistry, astronomy, and meteorology (Figure 1.1). Geology primarily deals with the science of earth. Chemistry deals with the structure, properties, and conversions of matter. Astronomy deals with the science of celestial bodies including planets, stars, and galaxies. Meteorology is the science of weather. It is believed that nature follows certain principles and there exists an order among a wide variety of natural phenomena. The primary role of physics is to *discover the fundamental principles of nature.* Physics holds the central position in the physical sciences, since it is concerned with the fundamental aspects of matter and energy as well as how they interact to make the universe function. The fundamentals of physics, on the one hand, include concepts and mathematics and, on the other hand, the practical applications of technology.

The field of physics is divided into the areas of mechanics, waves, optics, thermodynamics, and electricity, which make up the *classical* physics developed between the sixteenth and nineteenth centuries by Galileo, Newton, Faraday, Maxwell, and others. Physics also includes modern topics developed in the twentieth century, such as relativity, quantum physics, and quantum field theory, as well as their appropriate applications in condensed matter, atomic and nuclear phenomena, elementary particles, and astrophysics.

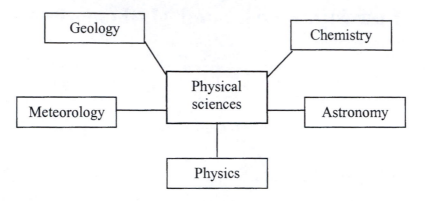

Figure 1.1

The Newtonian mechanics that holds perfectly well for ordinary moving objects was found to be inadequate for objects moving close to the speed of light and has been replaced by the special theory of relativity of Einstein (1905). For extremely small objects (comparable to the size of an atom or less), Newtonian mechanics has been superseded by quantum mechanics, as developed in the early twentieth century by Schrödinger, Heisenberg, and others. In the middle of the twentieth century, quantum field theory was developed that combined relativity and quantum principles. The great four areas of physics are shown in Figure 1.2.

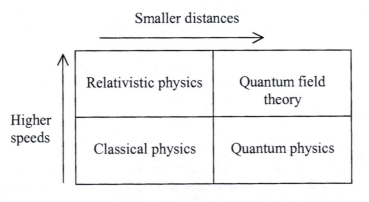

Figure 1.2

A solid foundation in physics is needed to understand any natural science. All the complexity and variety that you see in the world around you is a manifestation of a few fundamental principles of physics. It should be fun for you to know why the universe behaves the way it does. For many people, physics is a mechanical process of collecting facts. I do not think so. Physics, like art, poetry, and painting, is a creative activity that engages the emotion and the intellect. Physics is not a copy of nature, but a re-creation of nature by the act of discovery.

Physicists love physics because physics is beautiful. Physics shows us the elusiveness of the ultimate truth. Once you really understand physics, it will show you the same aesthetic qualities that you find in an exquisite work of art, a beautiful poem, or a pleasant piece of music.

The art of conversing with stone is called physics. The question-and-answer periods of the conversations are called experiments. ... The language spoken is mathematics. —J. T. Fraser

The Book of nature is written in mathematical characters. — Galileo Galilei

1.2   Mathematics: The Language of Physics

Physics is a fascinating science, and make no mistake about it—physics is not mathematics. *Mathematics is the study of numbers and the geometry of objects.* It is based on logic and the rules of counting. *Physics is the study of the rules that govern nature and of the fundamental processes occurring in nature.* Observation, careful experiments, and measurements comprise one aspect of physics. The other aspect of physics is the creation or invention of theories that explain the observation. Theoretical physics depends heavily on mathematics. You should find it very interesting that the principles and laws of physics can be expressed in terms of beautiful mathematical equations. Trying to study physics while totally avoiding mathematics is practically impossible. It would be the equivalent of studying Shakespeare while avoiding the written word. It simply cannot be done meaningfully.

People have noticed such a connection between physics and mathematics from ancient times. In ancient Greece, Pythagoras first noticed that pleasing notes were produced when the length of a string was changed by whole numbers relative to its original length. In the seventeenth century, Galileo discovered that the distance a freely falling object fell could be expressed by an equation that involves the square of time. The relationship between the cause (force) and effect (acceleration) for the motion of an object as expressed by Newton's second law of motion is a simple mathematical statement. Newton developed calculus (which was also developed by Leibniz), a powerful mathematical tool, to probe the changes in nature. For example, the position, velocity and acceleration of a mass attached to a spring at any time can be determined using calculus. In fact, all basic laws of physics can be expressed in beautiful and essentially simple mathematical forms. But why nature follows such beautiful mathematical equations is a mystery. I do not think anybody knows the answer to this question. But we are fortunate that nature has allowed us to explore its hidden truth and beauty through the use of mathematics developed by humans.

> *The most fundamental ideas of science are essentially simple, and, as a rule, can be expressed in a language comprehensible to every one.* —Albert Einstein

1.3   Why Does Physics Appear to Be Difficult?

It is not uncommon to hear from students that physics is difficult. There are some genuine reasons for this feeling:

- First, most physics courses rely heavily on mathematics for the understanding of physical processes. Real-world problem solving in physics also requires knowledge of mathematics. If you are anxious about your skills in mathematics, you will find physics to be difficult.

- Although the most fundamental ideas or concepts of physics are essentially simple, sometimes physics instruction is not always engaging.

- In introductory physics courses, you need to learn a great deal of course material within just two semesters. You must spend a lot of energy and effort to solve homework problems, read the text, attend laboratories, and take quizzes and exams. You might also need additional help from the instructor outside of the classroom. You may fail to get the additional help due to time conflicts in your busy schedule or the available hours of the instructor or teaching assistant.

- Finally, getting a good grade in a physics course is a challenge, even if you do attend classes and work regularly.

The point is how well you are prepared to meet the challenge. The obstacles may be several, but the primary factor that leads to success is your competence in using mathematics in physics with ease and accuracy. Meeting the math prerequisite for a physics course may not be enough. Students who have not recently taken a course that required mathematical skill often find themselves unable to keep pace in a physics course because of a lack of practice or familiarity with the necessary mathematical tools.

Do not worry about your difficulties in mathematics, I can assure you that mine are still greater. —Albert Einstein

1.4   Math Anxiety and How to Deal with It

Feeling nervous before an exam is not uncommon among students. Feeling slightly nervous when it comes to math is also not uncommon. However, if you really panic because you need to do math, it is likely that you have math anxiety. You need to ask yourself what are the causes of this anxiety and find out the root of these causes. Admit that math is sometimes not obvious and you, like others, need to work hard to understand. I have the following suggestions to reduce this anxiety:

- *Practice math to develop your math skills.* This is the first and most important advice for dealing with this anxiety. Once you have adequate math skills, you can concentrate more on the physics in a physics course. Talk with your physics instructor to find out what would help you to develop appropriate skills in mathematics.

- *Join a physics team involving three to four students* from the same physics class and *meet regularly* to discuss math as well as physics problems and concepts. You will find that you can develop your math skills and learn physics much easier in a group than on your own. Research has shown that students can learn well from their peers.

- Choose your teammates carefully to mix good work and fun. Choose a quiet and comfortable regular meeting place. Participate actively in the group and in the class. The more you discuss a problem with others, the easier it will be to recall. If you are lost, ask a team member to clearly explain. Do not copy from others.

- *Keep your body healthy.* Healthy bodies can think quicker and can work harder.

- *Take a break and relax if you are frustrated or nervous.* Learn and practice the techniques recommended for relaxing and beating stress. You will be able to concentrate more in math if you are less stressed.

- *Do not give up, get help.* There is no shame in getting help. If you need a remedial course in mathematics, take it. Many institutions offer a *physics recitation* course parallel to an introductory physics course. The physics recitation course is a good chance to practice problems and develop appropriate math skills.

- Remember, math anxiety can certainly be remedied if you make a serious effort. You can master mathematics if you work at it with caution and care.

When I really understand something, it is as if I had discovered it myself. —Richard Feynman

1.5   Developing Good Study Habits: Studying for Success

At the college level it is essential to study properly for success. First, the course material may be difficult to understand in one sitting. Second, you may not have enough time available between classes and extracurricular activities. Therefore, developing good study habits and proper study technique ise necessary steps for your success. Use the following tips to help you to develop proper study techniques and good study habits.

- *Plan your study wisely.* That is, make a realistic timetable that will include your study times as well as times for other activities that you need and want to do. The plan must be *realistic* so that you can stick to it throughout the course. *Maintain a calendar* to record important commitments such as tests, exams, due dates of papers, and so on.

- *Be a good reader.* Reading text is one critical part of the learning process, because all material that is included cannot be presented in the lecture. Use a highlighter to highlight the *main ideas*. Use a pencil to *take notes* in the margins. Always *think* about what you have read. Work the example problems of the section that you have read. Worked-out examples in the texts are usually helpful, and these may represent the standard of your test or exam.

- *Come to the class prepared.* That is, study the previous lecture and at least the outlines of the next lecture from the text before you attend the next class. Concentrate on the lecture. Concentration is a skill that can be improved by practice.

- *Interact with your instructor* in the class. Come to the class with questions from your reading. It is a good habit to sit in the class near the center of the front row so that instructor can clearly see your activities and you will have the minimum chance of being distracted.

- *Take good notes in class.* Your notes must include *main ideas* along with *supporting details* and *examples*. Rewrite your notes and create flash cards with important facts, laws, and vocabulary. Keep the flash cards handy and look at them as often as needed.

- *Join a study group* that can appropriately mix good work and fun. Choose your partners in the group carefully. Have discussion with others in the group and practice problems by collaborative efforts. Knowing or understanding a topic or a problem may not be enough; only *practice* can make you perfect in a test or exam.

- *Prepare well for the tests or exams, starting at least a week before.* Preparation is a *continuous daily activity* and not a task to be started the night before a test or an exam. As you study for the test or exam, ask yourself what questions or types of problems could be asked.

- Take a break from your study when needed. Also, take a break if it helps you increase your concentration on your studying. Do not feel guilty if you need a break.

- *Get help when you need it.* Meet your instructor outside of class to ask questions about concepts, walk through homework assignments, and review for tests and exams. Do not feel guilty if you need to see your instructor often during office hours. Discuss old tests, quizzes, exams, or practice problems with your instructor. Your library is a useful resource for books and other academic material. Take advantage of this. Practice old tests, quizzes, or exams, if available. Important study guide material or other resources for the course are often on reserve at the library. Use them as often as necessary.

- *Keep your body healthy and learn how to relax.* Stress can have negative effects on you and your thoughts. Too much stress can cause anxiety or even depression.

You do not know anything until you have practiced. —Richard Feynman

1.6   Solving Problems in Physics

To learn physics, you must practice by solving problems. Your ability to solve problems is one of the measures of whether you truly understand physics. To solve problems in physics, you must have the necessary mathematical skills. Remember that many problems can be solved by more than one method. Research has shown that students can solve problems well by collaborative efforts. Although there is no fixed recipe for solving problems in physics, you should find the following steps useful:

- *Read the problem carefully.* It is important to understand the meaning of each word in the problem. Recognize the meaning of different terms (e.g., *at rest* means that the initial velocity $v_0 = 0$; *a smooth, flat surface* means a horizontal and frictionless surface; moving at a *constant velocity* means that the acceleration $a = 0$). Read the entire problem. Texts often provide important hints at the end of a multiple part problem. If you only read part (a), you will miss the hints.

- *List data.* Some of the listed data may be redundant in some problems. For example, the mass of an object will not be necessary to determine its acceleration from the knowledge of time and its initial and final velocities.

- *Check units in the data and convert units where necessary.* Be careful about the units. Convert to Systéme International (SI) units whenever possible. Experience shows that students frequently get wrong results for problems either because the units are wrong or they made mistakes during unit conversions.

- *Draw a sketch or diagram of the problem wherever appropriate.* A simple sketch that depicts the spatial arrangements described in a problem greatly aids understanding of the problem. Some problems may require the adoption of a *sign convention* (e.g., upward direction as positive and downward direction as negative). *Stick with the same sign convention* for the whole problem. Some problems may require *setting up a convenient coordinate system* for the solution. In most situations, the Cartesian coordinate system (with horizontal direction as the x-axis and vertical direction as the y-axis) is suitable. However, for the motion of an object on an inclined plane the most convenient axes are usually the x-axis along the plane of inclined surface, and the y-axis perpendicular to the plane of the inclined surface.

- *Devise a strategy or plan for the solution.* Determine which physical principles and equations of physics can be applied. That is, think about the physics behind the problem and then relate the situation to physical principles that describe the physics (e.g., if the problem involves force and acceleration, then it involves Newton's second law; if the problem involves a collision between two objects, it relates to the conservation principle of linear momentum).

- *Translate principles into the language of equations.* For example, if you are dealing with a kinematics problem, and the initial and final velocities and time are given, use the equation $v = v_0 + at$.

- Discuss with others if the problem needs to be solved by collaborative efforts.

- *Substitute given quantities into equations,* and use your knowledge of mathematics to solve them or to do calculations to get the result. *Be aware of the common mathematical mistakes* (e.g., $1/a + 1/b$ is not equal to $1/(a + b)$; $x^2 + y^2$ is not equal to $(x + y)^2$).

- *Report the result in scientific notation keeping the proper number of significant figures with the proper unit.*

- *Evaluate your answer and determine whether the result makes sense* (e.g., find whether the result has the expected magnitude, dimension, unit, sign, etc.). In some situations, only one of the two mathematical solutions is physically acceptable. If you solve equations to get the result, substitute the results in the equations to verify whether they satisfy the equations. If the problem can be solved by an alternative method, you may also quickly check the result using the alternative method.

Example 1.6.1

Determine the time taken by a ball to roll with a constant velocity of 0.24 meters per second (m/s) across a 4.0-meter-long table.

The problem may seem simple, but the main purpose of this problem is to take you through the steps of good problem solving.

1. Read the problem.

Note that the phrase *constant velocity* means there will be no acceleration ($a = 0$).

2. List data.

$v = 0.24$ m/s
$x_i = 0.0$ m
$x_f = 4.0$ m
$a = 0$
$\Delta t = ?$ ← what we want to know

3. Check units.

In this case each unit is in standard form so no conversions are necessary.

4. Draw a sketch.

5. Devise a strategy or plan for the solution.	This is a kinematics problem that can be solved using the definition of constant velocity.

6. Translate principles

$$v = \frac{x_f - x_i}{\Delta t}$$

into the language of equations.

7. Substitute given quantities into equations and solve for unknown.

(i).   Insert numeric values:

$$0.24 \text{ m/s} = \frac{4.0 \text{ m} - 0.0 \text{ m}}{\Delta t}$$

(ii).   Multiply both sides by Δt:
       0.24 m/s $\Delta t = 4.0$ m.

(iii). Divide by 0.24 m/s:

$$\Delta t = \frac{4.0 \text{ m}}{0.24 \text{ m/s}} = 16.667 \text{ s}$$

8. Report the result in scientific notation with significant digits.

Use two significant digits, since all initial data are given to two significant digits:
$\Delta t = 1.7 \times 10^1$ s.

9. Evaluate your answer.

Units: Seconds—correct dimensions, since we were looking for a time.
Magnitude: Since traveling about 1/4 m/s and needed to travel 4 m, 17 s seems reasonable.
Sign: Positive, as expected.
Does the solution answer the question posed in problem? Yes.

Example 1.6.2

Draw a simple sketch to illustrate each of the following situations that you may find in a physics problem:

(a) A box resting on an inclined surface.

(b) A ball rolling down an incline.

c) An object is pulled at an angle of 45°
above the horizontal on a horizontal
surface.

(d) A ladder is leaning against a wall.

(e) An airplane flies 50 mi east and then
makes a left turn at an angle of 30° north
of east and flies another 100 mi.

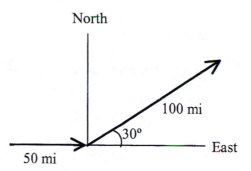

(f) Collision of two solid balls moving in the same direction before, during, and after
collision (moving in the opposite direction).

Chapter 2
Scientific Notation, Units, and Dimensions

2.1 Powers of 10 and Scientific Notation

In physics, you will deal with numbers that are extremely small and extremely large. For example, the lifetime of an unstable particle, called the muon, is 0.0000022 second, which is very small, whereas the mass of earth is 5,970,000,000,000,000, 000, 000,000 kilograms, which is very large. Such extremely small and large numbers take space and time to write. Scientists express such very small or large quantities in *scientific notation* using only the significant figures. Note the following notation:

$$10^0 = 1 \qquad\qquad 10^0 = 1$$

$$10^1 = 10 \qquad\qquad 10^{-1} = \frac{1}{10} = 0.1$$

$$10^2 = 10 \times 10 = 100 \qquad\qquad 10^{-2} = \frac{1}{10 \times 10} = 0.01$$

$$10^3 = 10 \times 10 \times 10 = 1000 \qquad\qquad 10^{-3} = \frac{1}{10 \times 10 \times 10} = 0.001$$

and so on.

In *scientific notation*, a number is written as the product of a number from 1 through 10 and an integer power of 10. That is, written in the scientific notation, every number has the following form:

$$(\text{A decimal between 1 and 10}) \times 10^{\text{An integer exponent}}$$

Thus, expressed in scientific notation,
the lifetime of muon = 2.2×10^{-6} s
the mass of earth = 5.97×10^{24} kg

Exercise
1. Express the following numbers in scientific notation:
    (a) 272000   (b) 0.000000534   (c) 324×10^2 (d) 128×10^{-6}
    (e) 0.0000054

2.2   Significant Figures and Rounding Off

The numbers dealt with in mathematics are exact. For example, when a mathematician writes 1, it means 1.000…. That is, 1 means exactly 1. The situation is different in physics. Many of the numbers dealt with in physics come from the measurements of physical quantities. The measured numbers of physical quantities are usually not exact. Exact numbers do not have any uncertainty or error. Measured numbers usually have some degree of uncertainty or error. For example, suppose the measured distance by a centimeter ruler is 2.53 cm. Here, 3 is an estimate meaning that the distance is not less than 2.5 cm or more than 2.6 cm. The 2, the 5, and the 3 in the measurement are called *significant figures*, because they each provide reliable information about the length of the object. A centimeter ruler measurement is good to three significant figures, whereas a precise micrometer can yield a more precise value of the measured length, although that will still not be exact. The number of *significant figures* in a physical quantity is equal to the number of digits that are estimated or known with some reliability. Also, in numerical calculations using a calculator, the calculator may give a result that has many figures, which are not meaningful.

Use the following rules to determine the significant figures:

- Zeros at the beginning of a number are not significant. For example, in 00056, the zeros are not significant.

- Zeros within a number are significant. For example, in 5.04, the zero is significant.

- Zeros at the end of a number after the decimal point are significant. For example, in 54.00, the zeros are significant.

In whole numbers without a decimal point that end in one or more zeros (for example, 500 kg), the zeros may or may not be significant. Scientific notation allows the number of significant figures to be expressed clearly. For example, consider the number 23,400. If the number is known up to three significant figures, we write 2.34×10^4, whereas if the number is known up to four significant figures, we write 2.340×10^4.

> Use the following rules to round off numbers:
>
> 1. If the first digit to be dropped is less than 5, leave the preceding digit as is (e.g., 2.162 becomes 2.16 rounded to three significant figures).
>
> 2. If the first digit to be dropped is 5 or greater, increase the preceding digit by one (e.g., 2.167 becomes 2.17 rounded to three significant figures).

Example 2.2.1

Express the following numbers in scientific notation with three significant figures:

   (a) 0.0241   (b) 79,241   (c) 0.000452

The numbers in scientific notation with three significant figures are as follows:

   (a) $0.0241 = 2.41 \times 10^{-2}$

   (b) $79,241 = 7.92 \times 10^{4}$ (*Note*: 7.9241 has been rounded off to 7.92.)

   (c) $0.000452 = 4.52 \times 10^{-4}$

Example 2.2.2

Express the following numbers as their decimal equivalents:

   (a) 49.6×10^{-2} (b) -6.50×10^{-4}

   (c) 0.4291×10^{3} (d) 1.21×10^{0}

The decimal equivalents are as follows:

   (a) $49.6 \times 10^{-2} = 0.496$ (by moving the decimal point two places left)

   (b) $-6.50 \times 10^{-4} = -0.000650$ (by moving the decimal point four places left)

   (c) $0.4291 \times 10^{3} = 429.1$ (by moving the decimal point three places right)

   (d) $1.21 \times 10^{0} = 1.21$

Example 2.2.3

The nearest star outside our solar system is Alpha Centauri, whose distance is approximately 25,700,000,000,000 miles. Express the result in scientific notation with three significant figures and four significant figures.

25,700,000,000,000 miles

We know that 25,700,000,000,000 miles = 2.57×10^{13} miles, written in scientific notation with three significant figures. Expressing with four significant figures, we find that the result is 2.570×10^{13} miles. Note that, in the latter case, the zero after the decimal point at the end of 2.570 is significant.

Example 2.2.4

In each of the following expressions, rewrite each factor and term in scientific notation and express the final result of the computation in scientific notation with three significant figures:

(a) 51.5×0.0000521

First, express each number in scientific notation.

51.5×0.0000521

$= (5.15 \times 10^{1}) \times (5.21 \times 10^{-5}) = (5.15 \times 5.21) \times 10^{-4} = 26.8315 \times 10^{-4}$

The result is not correctly written in scientific notation. To get the decimal place in the right place, we move one decimal place to the left, increasing the exponent by + 1. Thus, the answer in scientific notation is 2.68315×10^{-3}. Keeping the result up to the three significant figures, we round off the result and rewrite it as 2.68×10^{-3}.

(b) $\dfrac{2348}{83.1 - 91.8}$

First, express each number in the fraction in scientific notation.

$$\frac{2348}{83.1 - 91.8} = \frac{2.348 \times 10^{3}}{8.31 \times 10^{1} - 9.18 \times 10^{1}} = -2.70 \times 10^{2}$$

Here the result is shown in scientific notation up to the three significant figures.

The result of a measured quantity, or of a problem, should be reported with the proper number of significant figures. Use the rules (mentioned in the following boxes) to determine the number of significant figures to be kept after adding, subtracting, multiplying, and dividing numbers.

When adding or subtracting numbers, leave the same number of decimal places in the answer as there is in the number with the least number of significant figures.

Add the lengths of two tables of lengths 6 ft and 4 ft, 3 in. Since the first table does not specify the inches, we must round our final answer to 10 ft. This is because, if no inches are specified, "6 ft" means "anywhere between 5 ft, 6 in to 6 ft, 6 in."

For example, suppose you need to add 702.3, 124.0, 42.06, and 0.0542. The result from your calculator is 868.4142. Now, look at your data and observe that in two of the numbers being added it is not clear what the figure in the second decimal place is. The result must be rounded off to one decimal place by dropping 142. Hence, the answer with proper numbers of significant figures is 868.4.

When multiplying and dividing numbers, leave as many significant figures in the answer as there are in the number with the fewest number of significant figures.

Usually, multiplication and division produce many more figures that are significant thus, the result must be rounded off. For example, suppose the measured length and width of a plate are 2.64 cm and 1.18 cm, respectively. Therefore, the area = 2.64 cm X 1.18 cm = 3.1152 cm^2. The result appears to have five significant figures, although the original quantities have three each. Therefore, the result can have only three significant figures and is 3.12 cm^2 (rounded to the second decimal place).

The next example also illustrates how to determine the correct number of significant figures in addition, subtraction, multiplication, and division.

Example 2.2.5
Determine the correct number of significant figures in the following operations:

(a) $(5.32 \times 10^{-4}) + (7.4 \times 10^{-4})$ (b) $(3.42 \times 10^{-6}) - (2.8 \times 10^{-6})$

(c) $(5.212 \times 10^{3}) \times (7.8 \times 10^{5})$ (d) $(8.2 \times 10^{5}) \div (7.62 \times 10^{2})$

(a) $(5.32 \times 10^{-4}) + (7.4 \times 10^{-4}) = (5.32 + 7.4) \times 10^{-4} = 12.72 \times 10^{-4}$

The final answer can have only the same number of decimal places in the answer as there are in the number with the least number of decimal places. The number with least number of decimal places in this problem is 7.4×10^{-4}. Since 7.4×10^{-4} is expressed to the one decimal place, the result is 12.7×10^{-4}. To express the result in scientific notation, move the decimal point to the left one position, increasing the exponent by +1. Thus, the result is 1.27×10^{-3}.

(b) $(3.42 \times 10^{-6}) - (2.8 \times 10^{-6}) = (3.42 - 2.8) \times 10^{-6} = 1.62 \times 10^{-6}$
The number with the least number of decimal places is 2.8×10^{-6}, which has two significant figures. The final answer can have only two significant figures and, hence, the result is 1.6×10^{-6}.

(c) $(5.212 \times 10^{3}) \times (7.8 \times 10^{5}) = (5212 \times 780000) = 4065360000 = 4.1 \times 10^{9}$, rounded

up to two significant figures

(d) $\dfrac{8.2 \times 10^{5}}{7.62 \times 10^{2}} = \dfrac{820000}{762} = 1076.1155 = 1.1 \times 10^{3}$, rounded up to two significant

figures

Exercises
2. (a) Express 200 kg in scientific notation with two significant figures.

   (b) Express 200 kg in scientific notation with three significant figures.

3. Round off the results in scientific notation to the proper number of significant figures:

    (a) Multiplication: $2.4 \text{ m} \times 3.65 \text{ m}$ (b)   Division: $\dfrac{7.24 \text{ m}}{1.116 \text{ s}}$

    (c) Addition: $8.1 \times 10^{4} \text{ m} + 9.1 \times 10^{3} \text{ m} + 0.008 \times 10^{6} \text{ m}$

2.3   Physical Quantity, Units, and Prefixes

In physics, you will deal with different physical quantities, such as length, mass, force, and others. A number and a unit express a *physical quantity*. There are two types of physical quantities: *base quantities* (or fundamental quantities) and *derived quantities*. Derived quantities can be expressed in terms of base quantities. In your study of physics, you will come across many different units. Some are familiar to you (such as inches, gram, etc.) and some may be unfamiliar (such as newtons, pascals, etc.). Use the *metric system* of units whenever possible in solving problems. In the metric system, you multiply (or divide) by a power of 10 when you change from a larger (or smaller) unit to a smaller (or larger) unit.

Table 2.1: International System (SI) of Units for Base Quantities

Quantity	Unit	Abbreviation
Length	meter	m
Mass	kilogram	kg
Time	second	s
Electric current	ampere	A
Temperature	kelvin	K
Luminous intensity	candela	cd
Amount of substance	mole	mol

Table 2.2: International System (SI) of Units for Derived Quantities

Quantity	Unit	Abbreviation
Angle	radian	rad
Solid angle	steradian	sr
Area	square meter	m^2
Volume	cubic meter	m^3
Density	kilogram per cubic meter	kg/m^3
Velocity	meter per second	m/s
Acceleration	meter per second squared	m/s^2
Angular velocity	radian per second	rad/s
Angular acceleration	radian per second squared	rad/s^2
Force	newton	$N\ (kg \cdot m/s^2)$
Torque	newton-meter	N·m
Work, energy, heat	joule	$J\ (N \cdot m)$
Power, radiant flux	watt	W
Pressure, stress	pascal	$Pa\ (N/m^2)$
Frequency	hertz	Hz (1/s)
Electric charge	coulomb	$C\ (A \cdot s)$
Voltage, electromotive force	volt	V (W/A)
Electric field	volt per meter	V/m
Electric resistance	ohm	Ω (V/A)
Capacitance	farad	$F\ (A \cdot s/V)$
Magnetic flux	weber	$Wb\ (V \cdot s)$
Inductance	henry	$H\ (V \cdot s/A)$
Magnetic flux density	tesla	$T\ (Wb/m^2)$
Luminous flux	lumen	Lm
Luminance	lux	lx
Activity (radioactive)	becquerel	$Bq\ (s^{-1})$
Absorbed dose (of radiation)	gray	Gy (J/kg)
Dose equivalent (dose equivalent index)	sievert	Sv (J/kg)

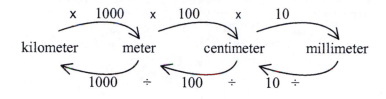

The SI unit (International System of Units) was adopted in 1960 as standard units to express physical quantities. The SI units of length, mass (i.e., the amount of matter), and time are *meter* (m), *kilogram* (kg), and *second* (s), respectively. These units are also called MKS units. The SI units of base and derived quantities are shown in Tables 2.1 and 2.2, respectively.

Prefixes

In the metric system, the larger and smaller units are defined in certain powers of 10 from the standard unit to make the conversion easy. The *prefixes* can be used to express units of any physical quantity in the metric system. To use computer memory, the common prefixes are kilobyte, megabyte, gigabyte, and terabyte. The prefixes, their symbols, and their meanings are listed in Table 2.3. Table 2.4 lists the most common prefixes used in introductory physics.

Table 2.3: Prefixes and Their Meanings

Prefix	Symbol	Meaning
peta	P	10^{15} = 1000 000 000 000 000
tera	T	10^{12} = 1 000 000 000 000
giga	G	10^{9} = 1 000 000 000
mega	M	10^{6} = 1 000 000
kilo	k	10^{3} = 1 000
deci	d	10^{-1} = 0.1
centi	c	10^{-2} = 0.01
milli	m	10^{-3} = 0.001
micro	μ	10^{-6} = 0.000 001
nano	n	10^{-9} = 0.000 000 001
pico	p	10^{-12} = 0.000 000 000 001
femto	f	10^{-15} = 0.000 000 000 000 001
atto	a	10^{-18} = 0.000 000 000 000 000 001

Table 2.4: Most Common Prefixes Used in Introductory Physics

Mass	kg (kilogram = 10^3 grams)
Length	mm (millimeter = 10^{-3} m), cm (centimeter = 10^{-2} m), km (kilometer = 10^3 m), μm (micrometer = 10^{-6} m), nm (nanometer = 10^{-9} meter)
Time	ms (millisecond = 10^{-3} s), μs (microsecond, 10^{-6} s)
Power	MW (megawatt = 10^6 W)
Frequency	kHz (kilohertz = 10^3 Hz), MHz (megahertz = 10^6 Hz)
Electric charge	μC (microcoulomb = 10^{-6} C)
Capacitance	μF (microfarad = 10^{-6} F), pF (picofarad = 10^{-12} F)
Electric current	mA (milliampere = 10^{-3} A), μA (microampere = 10^{-6} A)
Resistance	kΩ (kilo ohm = 10^3 Ω), MΩ (mega ohm = 10^6 Ω)

Example 2.3.1

Write the following as decimal numbers with standard units:

(a) 81 micrometers (b) 32 picoseconds (c) 11.2 femtometers (d) 2.5 gigahertz

(a) 81×10^{-6} m = 0.000081 m
(b) 32 picoseconds = 32×10^{-12} s = 0.000000000032 s
(c) 11.2 femtometers = 11.2×10^{-15} m = 0.0000000000000112 m
(d) 2.5 gigahertz = 2.5×10^9 Hz = 2500000000 Hz

Exercise

4. Rewrite the following quantities using prefixes.
    Wavelength of helium-neon laser = 0.0000006328 m
    Size of a virus = 0.0000001 m

    The radius of the nucleus of a carbon [$^{14}_{6}$C] atom = 0.0000000000000029 m

    Rest energy of an electron = 511,000 electron volts (eV)

2.4   Unit Conversions

Units can be converted from one system of units to another by multiplying by the appropriate factors, called conversion factors. *Conversion factors* are simply equivalence statements expressed in the form of ratios. For example, 1 in. = 0.0254 m, where the ratio is 0.0254. Some important conversions are listed in Appendix A. In unit conversion, multiply or divide by the appropriate conversion factor so that the unwanted units cancel, thus leaving the desired unit. This is shown in the next example.

Example 2.4.1

The posted speed limit on a highway is 65 mi/h. Determine the speed limit in kilometers per hour (km/h) and in meters per second (m/s).

Speed limit = 65 mi/h

1 mi = 1.609 km   That is, $1 = \left(1.609\dfrac{km}{mi}\right)$.

Therefore, to convert mi/h to km/h, multiply by $\left(1.609\dfrac{km}{mi}\right)$.

Thus, $65\dfrac{mi}{h} = \left(65\dfrac{\cancel{mi}}{h}\right)\left(1.609\dfrac{km}{\cancel{mi}}\right) = 105\dfrac{km}{h}$.

Also, 1 mi = 1609 m and 1 h = 3600 s. That is, $1 = \left(1609\dfrac{m}{mi}\right)$ and $1 = \left(\dfrac{1h}{3600\,s}\right)$.

Therefore, $65\dfrac{mi}{h} = \left(65\dfrac{\cancel{mi}}{\cancel{h}}\right)\left(1609\dfrac{m}{\cancel{mi}}\right)\left(\dfrac{1\cancel{h}}{3600\,s}\right) = 29\dfrac{m}{s}$.

Example 2.4.2

Atmospheric pressure is 14.7 lb/in.^{2}. Express atmospheric pressure in newtons per square meters (N/m^2), using the conversions 1 in. = 0.0254 m and 1 N = 0.225 lb.

1 N = 0.225 lb and 1 in. = 0.0254 m.   That is, $1 = \left(\dfrac{1N}{0.225\,lb}\right)$ and $1 = \left(\dfrac{1\,in.}{0.0254\,m}\right)$.

Therefore, atmospheric pressure $= 14.7\dfrac{lb}{in.^2} = \left(14.7\dfrac{lb}{in.^2}\right)\left(\dfrac{1\,N}{0.225\,lb}\right)\left(\dfrac{1\,in.}{0.0254\,m}\right)^2$

$$= 1.01 \times 10^5\,\dfrac{N}{m^2}.$$

Exercises

5. The highest waterfall is Angel Falls, whose height is 979.0 m. Express its height in miles.

6. The age of the universe is believed to be 15 billion years. Express the result in seconds.

7. A *light year* is the distance traveled by light in one year. Light travels 3.0×10^8 m in 1 second. An *astronomical unit* (AU) is the average distance from the earth to the sun, which is 1.5×10^8 km. Express 1 light year in terms of AU.

2.5   Dimension in Physics

The *dimension* of a physical quantity represents the *type* of the quantity, regardless of its units. The dimensions of length, mass, and time are expressed as [L], [M], and [T], respectively. For example, speed for a moving object, which is defined as distance over time, has the dimension [L]/[T].

Any valid mathematical equation must be dimensionally consistent. In deriving formulas or solving problems, dimension checks help verify each step of the derivation or solution. Each term in the equation must have the same dimensions.

Take an example that illustrates how a dimension check helps. The rate at which a pendulum moves back and forth depends on the length of the pendulum and the local acceleration due to gravity. The number of complete oscillations per unit time is called the frequency, f. Suppose you do not remember whether the frequency f of a pendulum is $f = \dfrac{1}{2\pi}\sqrt{\dfrac{l}{g}}$ or $\dfrac{1}{2\pi}\sqrt{\dfrac{g}{l}}$. Here, l is the length of the pendulum, whose dimension is [L], and g is called the acceleration due to gravity whose dimension is $[L]/[T^2]$.

The dimension of the right-hand side of the first formula is

$$\sqrt{\frac{[L]}{[L/T^2]}} = \sqrt{[T^2]} = [T]$$

The dimension of the right-hand side of the second formula is

$$\sqrt{\frac{[L/T^2]}{[L]}} = \sqrt{\frac{1}{[T^2]}} = \frac{1}{[T]}$$

Now, frequency being the number (of oscillations) in unit time must have the dimensionn of reciprocal of time, 1/[T]. Obviously, the second equation is correct.

Important Note: A dimension check is not a guarantee that a formula or an answer is correct, because the dimensionless factors (such as 2π, ½, etc.) do not show up in the dimension check. But the answer or the solution will not be correct, if the dimensions are not correct.

8. The speed of an object is given by the equation $v = At^2 + Bt$, where t represents time. Determine the dimensions of A and B.

9. Show that the equations (i) $x = vt + \dfrac{1}{2}at^2$ and (ii) $v^2 = v_0^2 + 2ax$ are dimensionally correct, where the dimensions of x, v, v_0, a, and t are [L], [L]/[T], [L]/[T], [L]/[T^2], and [T], respectively.

10. According the Kepler third law of planetary motion, the square of the period (the time taken by a planet to complete one revolution) is given by $T^2 = (4\pi^2/GM_S)\, r^3$, where T is the period, M_S is the mass of the sun, r is the mean distance of the planet from the sun, and G is a constant called the universal gravitational constant. Determine the dimension of G.

11. If the dimension of the energy E is [M][Ll]^2/[T]^2, which of the following equations is dimensionally correct—$E = mc$, $E = mc^2$, and $E = mc^3$, if m represents the mass of an object and c represents the speed of light.

Answers to Exercises in Chapter 2

1. (a) 2.72×10^5

    (b) 5.34×10^{-7}

    (c) 3.24×10^4

    (d) 1.28×10^{-4}

    (e) 5.4×10^{-6}

2. (a) 2.0×10^2 kg

    (b) 2.00×10^2 kg

3. (a) $8.8 \times 10^0 \, m^2$

    (b) 6.49 m/s

    (c) 9.8×10^4 m

4. 632. 8 nm, 0.1 μm, 29 fm, 0.511 MeV

5. 0.608 miles

6. 4.7×10^{17} s

7. 1 light year = 6.3×10^4 AU

8. $[L]/[T^3]$, $[L]/[T^2]$

10. $[L]^3/[M][T]^3$

11. $E = mc^2$

Chapter 3
Algebra:Its Rules and Solving Equations

Introduction

Mathematics is the study of numbers and the geometry of objects. It is a tool to study nature and its changes. Algebra is essentially a system of rules and procedures for dealing with numbers and logically exploring the relationships between variables using various symbols. In physics, we deal with numbers, and one of our major concerns is to describe interdependences among physical quantities with equations. Algebra is very useful to explore such interdependences. For example, when an apple falls from a tree, its vertical distance (y) of fall can be described by an equation $y = Ct^2$, where C is a constant that is related to the gravity of the earth and t represents time.

3.1   Symbols and Letters

Different symbols and letters are used in algebra and, in general, in any mathematics to mean or abbreviate different concepts or operations. Some of the symbols and letters used in algebra, together with their meanings, are listed in Table 3.1. Greek letters are often used in mathematics (including algebra) and physics. The letters are shown in Table 3.2.

Table 3.1: Symbols and Letters

SYMBOL	MEANING	SYMBOL	MEANING
$=$	Equal to	$\therefore$	Therefore
$\approx$	Approximately equal to	Σ	Sum
$\sim$	Same order as	Π	Product
$>$	Greater than	$\sqrt{}$	Square root
$\geq$	Greater than or equal to	$x!$	Factorial of x
$<$	Less than	$\neq$	Not equal to
$\leq$	Less than or equal to	∞	Infinity
$>>$	Much greater than	$\propto$	Proportional to
$<<$	Much less than	$\pm$	Plus or minus
$\equiv$	Equivalent to	$\mp$	Minus or plus
$\Rightarrow$	Implies	Δx	Change in x
$/$	Divided by		

Table 3.2: Greek Letters

	Uppercase	Lowercase		Uppercase	Lowercase
Alpha	A	α	Nu	N	ν
Beta	B	β	Xi	Ξ	ξ
Gamma	Γ	γ	Omicron	O	o
Delta	Δ	δ	Pi	Π	π
Epsilon	E	ε	Rho	P	ρ
Zeta	Z	ζ	Sigma	Σ	σ
Eta	H	η	Tau	T	τ
Theta	Θ	θ	Upsilon	Y	υ
Iota	I	ι	Phi	Φ	ϕ
Kappa	K	κ	Chi	X	χ
Lambda	Λ	λ	Psi	Ψ	ψ
Mu	M	μ	Omega	Ω	ω

Dealing with Numbers

3.2   Numbers, Fractions, and Basic Rules

A *fraction* is the ratio of two numbers, a and b, and is expressed as a/b or $\dfrac{a}{b}$ (where $b \neq 0$). The numbers a and b are called the *numerator* and the *denominator* of the fraction, respectively.

The numbers ..., –4, –3, –2, –1, 0, 1, 2, 3, 4, ... are known as *integers*. Numbers that can be expressed as a fraction with an integer in its numerator and a nonzero integer in its denominator are called *rational numbers*. A rational number can also be expressed in decimal notation. For example, 10½ is a rational number, which, in decimal notation, is 10.5. *Irrational numbers* are numbers that cannot be expressed as a ratio of two integers. Their decimal notations do not terminate or repeat. Some common irrational numbers that appear in physics are shown in Table 3.3. Rational and irrational numbers together constitute what are called *real numbers*.

Table 3.3: Common Irrational Numbers

$\pi = 3.141592...$	$e = 2.718282...$	$\sqrt{2} = 1.414213...$
$\sqrt{3} = 1.732050...$	$1/\sqrt{2} = 0.707106...$	$\sqrt{3}/2 = 0.866025...$
$\log 2 = 0.301029...$	$\ln 2 = 0.693147...$	$\ln 10 = 2.302585...$

Addition, Subtraction, Multiplication, and Division of Real Numbers

The following properties of real numbers are useful when you add or multiply two or more real numbers, or add 0 to a real number, or multiply a real number by 1.

For all real numbers a, b, and c,

(i) $(a + b)$, $(a - b)$, $a \cdot b$ and a/b $(b \neq 0)$ are real numbers (Closure properties)

(ii) $a + b = b + a$ and $a \cdot b = b \cdot a$ (Commutative properties)

    Example: $2 + 7 = 9 = 7 + 2,$ $3 \cdot 6 = 18 = 6 \cdot 3$

(iii) $(a + b) + c = a + (b + c)$, $(ab)c = a(bc) = b(ac)$

    and $(ab/c) = a(b/c) = b(a/c)$ (Associative properties)

    Example: $2 + 5 + 9 = (2 + 5) + 9 = 16 = 2 + 14 = 2 + (5 + 9)$

              $2 \cdot 5 \cdot 4 = 2(5 \cdot 4) = 2 \cdot 20 = 40 = 10 \cdot 4 = (2 \cdot 5)4$

(iv) $a(b + c) = ab + ac$ (Distributive property)

    Example: $2(3 + 4) = 2 \cdot 7 = 14 = 6 + 8 = 2 \cdot 3 + 2 \cdot 4$

(v) $a + (-a) = 0$ and $a \cdot (1/a) = 1$ (Inverse properties)

(vi) $a + 0 = a$ and $a \cdot 1 = a$ (Identity properties)

Rules for Combining Signs

When multiplying or dividing numbers of like sign, the result is positive; otherwise the result is negative. For example,
$(+a)(+b) = (-a)(-b) = +ab$, and $(+a)(-b) = (-a)(+b) = -ab$.

Addition, Subtraction, Multiplication, and Division of Fractions

Use the following rules when adding, subtracting, multiplying, and dividing tw fractions.

Addition and Subtraction $\dfrac{a}{b} \pm \dfrac{c}{d} = \dfrac{ad \pm bc}{bd}$

Multiplication $\dfrac{a}{b} \cdot \dfrac{c}{d} = \dfrac{ac}{bd}$

Division $\dfrac{\dfrac{a}{b}}{\dfrac{c}{d}} = \dfrac{ad}{bc}$

Note: *Be careful of mistakes when you add two fractions.* For example,

$$\frac{a}{b} \pm \frac{c}{d} \text{ is not equal to } \frac{a \pm c}{b \pm d}. \qquad \text{Also, } \frac{1}{a} \pm \frac{1}{b} \text{ is not equal to } \frac{1}{a \pm b}.$$

In fact, $\dfrac{1}{a} \pm \dfrac{1}{b} = \dfrac{b}{ab} \pm \dfrac{a}{ba} = \dfrac{b \pm a}{ab}$.

Important Note: Find a common denominator when you add two or more fractions. In the preceding example, the common denominator is *ab*, which is obtained by multiplying the numerator and denominator of the first and second fractions by *b* and *a*, respectively. The next example will illustrate an application of this rule in physics.

Example 3.2.1 (from electricity)

The ability of an electrical component to oppose the flow of electricity through it is called its *electrical resistance R*. The *equivalent electrical resistance R_{eq}* of two electric components (such as two electric lightbulbs) R_1 and R_2 connected in a parallel combination is given by the formula $1/R_{eq} = 1/R_1 + 1/R_2$. Find the equivalent resistance of two resistances $R_1 = 100.0$ ohm and $R_2 = 200.0$ ohm. (Ohm is the unit of electric resistance.)

We have

$$a = R_1 = 100.0 \text{ ohm}$$
$$b = R_2 = 200.0 \text{ ohm}$$

This problem uses the rule of addition.

Parallel combination of R_1 and R_2

Applying this rule, we get

$$\frac{1}{R_{eq}} = \frac{1}{R_1} + \frac{1}{R_2} = \frac{R_1 + R_2}{R_1 \cdot R_2}$$

Substitute the values of R_1 and R_2.

$$= \frac{100.0 \text{ ohm} + 200.0 \text{ ohm}}{100.0 \text{ ohm} \cdot 200.0 \text{ ohm}}$$

Thus,

$$R_{eq} = \frac{100.0 \text{ ohm} \cdot 200.0 \text{ ohm}}{100.0 \text{ ohm} + 200.0 \text{ ohm}}$$

$$= \frac{20000}{300} \text{ ohm} = 66.7 \text{ ohm}$$

Exercise

1. The equivalent resistance of three electrical resistances connected in parallel is given by

$$\frac{1}{R_{eq}} = \frac{1}{R_1} + \frac{1}{R_2} + \frac{1}{R_3}. \text{ Show that } R_{eq} = \frac{R_1 R_2 R_3}{R_2 R_3 + R_3 R_1 + R_1 R_2}.$$

Division with Zero

$\frac{15}{3} = 5$, since $5 \cdot 3 = 15$. Thus, $\frac{0}{x} = 0$, since $0 \cdot x = 0$. But $\frac{x}{0}$ is undefined, since no number multiplied by 0 gives x. Also, since any number multiplied by 0 gives 0, $\frac{0}{0}$ is indeterminate.

Important Note for Division with Zero	
$\frac{x}{0}$	Undefined
$\frac{0}{0}$	Indeterminate
$\frac{0}{x}$	0, if $x \neq 0$

Percent

A *percent* is the numerator of a fraction whose denominator is always 100. In other words, *percent* means *per hundred*. Thus, $20\% = \frac{20}{100} = 0.20$.

In problems involving percentages, the word *of* usually indicates multiplication. For example,

$$20\% \text{ of } 6000 = \frac{20}{100} \times 6000 = 1200$$

In this example, 20% of 6000 = 1200, the percent is called the *rate* (r), 6000 is called the *base* (b) and 1200 is called the *percentage* (p). For any percentage problem,

$$\text{rate} \cdot \text{base} = \text{percentage} \quad \text{or} \quad r \cdot b = p$$

Example 3.2.2

After examining 200 shirts, a quality control inspector found 10 with defective stitching, 6 with mismatched designs, and 2 with an incorrect label. What percent was defective?

Defective shirts = 10 + 6 + 2 = 18. Here, $b = 200$ and $p = 18$.

Use the relation	$r \cdot b = p$
Substitute the values.	$r \cdot 200 = 18$
Divide by 200 to solve for r.	$r = \dfrac{18}{200} = 0.09$
Multiply by 100 to get the percentage.	$r = 0.09 \times 100\% = 9\%$

3.3   The Number Line and Absolute Value

The following number line represents real numbers. Numbers to the *right* of 0 are called *positive* and numbers to the *left* of 0 are called *negative*. The number zero is neither positive nor negative.

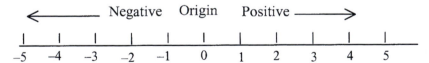

The magnitude of the distance between a number x and 0 is called the *absolute value* of x and is written as $|x|$. For example, since the distance between 4 and 0 is 4 units and the distance between –4 and 0 is also 4 units, $|4| = 4$ and $|-4| = 4$. Obviously, the absolute value of any number is either positive or zero.

The absolute value of a quantity is needed in physics, where the magnitude of the quantity is important, but not its sign. For example, when calculating the average deviation from the average result of an experimental variable, first calculate the absolute deviation of individual results $|x_1 - \bar{x}|$, $|x_2 - \bar{x}|$, ... (where x_1, x_2, ... are the results of individual measurements and $\bar{x}$ is the average value of the measurements).

Then take the average by dividing the absolute deviations by the number of measurements. Thus, if there are N individual measurements, then

$$\text{Average deviation} = \frac{\left|x_1 - \bar{x}\right| + \left|x_2 - \bar{x}\right| + ... + \left|x_N - \bar{x}\right|}{N}$$

Example 3.3.1

(a) Determine the absolute value of 3, –8, and 0.

(b) The individual results of length measurements in a physics experiment are 4.0 cm, 3.9 cm, 3.7 cm, 4.1 cm, and 3.8 cm. Determine the absolute deviations of the individual results from the mean. Determine the average deviation.

(a) $\left|3\right| = 3$ $\qquad\qquad$ $\left|-8\right| = 8$ $\qquad$ $\left|0\right| = 0$

(b) The mean $= (4.0 \text{ cm} + 3.9 \text{ cm} + 3.7 \text{ cm} + 4.1 \text{ cm} + 3.8 \text{ cm})/5 = 3.9 \text{ cm}$.

The absolute deviations of the individual results are as follows:

$\left|4.0 \text{ cm} - 3.9 \text{ cm}\right| = \left|0.1 \text{ cm}\right| = 0.1 \text{cm},$ $\qquad$ $\left|3.9 \text{ cm} - 3.9 \text{ cm}\right| = \left|0.0 \text{ cm}\right| = 0.0 \text{ cm},$

$\left|3.7 \text{ cm} - 3.9 \text{ cm}\right| = \left|-0.2 \text{ cm}\right| = 0.2 \text{cm},$ $\qquad$ $\left|4.1 \text{ cm} - 3.9 \text{ cm}\right| = \left|0.2 \text{ cm}\right| = 0.2 \text{ cm},$

$\left|3.8 \text{ cm} - 3.9 \text{ cm}\right| = \left|-0.1 \text{ cm}\right| = 0.1 \text{cm}$

Average deviation $= (0.1 \text{ cm} + 0.0 \text{ cm} + 0.2 \text{ cm} + 0.2 \text{ cm} + 0.1 \text{ cm})/5 = 0.12 \text{ cm}$

Proportion, Ratio, and Relationship between Quantities

3.4   Proportion and Ratio

The essence of physics is to describe and verify the relationships among physical quantities. The relationships are often simple: Two quantities may be directly proportional to each other, inversely proportional to each other, or one may be proportional to the square of the other.

Two quantities are said to be *directly proportional* to one another if an increase (or decrease) of the first quantity causes an increase (or decrease) of the second quantity by the same factor. If y is directly proportional to x, the direct proportionality is written as

$$y \propto x$$

The ratio y/x is a constant, say, k. That is, $y_1/x_1 = y_2/x_2 = k$.

Thus, in a direct proportion, the relationship is described by an equation, $y = kx$.

Look at the following data for the diameter and the circumference of a circle.

Diameter, d (m)	Circumference, C (m)
0.240	0.754
0.420	1.319
0.720	2.262

The data show that the ratio of the circumference C to the diameter d is always 3.14, which is called π. We can say that the circumference of a circle is directly proportional to its diameter, and $C = \pi d$, where π is the constant of proportionality.

In general, in a direct proportion, $\dfrac{a}{b} = \dfrac{c}{d}$, a and d are called the *extremes* and b and c are called the *means*. Multiplying both sides by bd, we find that

$$b d \cdot \frac{a}{b} = b d \cdot \frac{c}{d} \qquad\qquad \text{or} \qquad a \cdot d = b \cdot c$$

That is, the product of the extremes is equal to the product of the means. This fact is useful in mathematics in solving for an unknown variable, say, x, from an equation, such as

$$\frac{x}{4} = \frac{21}{7}$$

Thus, $7x = 4 \cdot 21$ or $x = \frac{4 \times 21}{7} = 12$

Example 3.4.1

If y is directly proportional to x, and $x = 2$ when $y = 16$, what is y when $x = 6$?

Since y is directly proportional to x, $y_1/x_1 = y_2/x_2 = k$.

Substitute the values. $\dfrac{y}{6} = \dfrac{16}{2}$

Multiply by $2 \cdot 6$. $2y = 16 \cdot 6 = 96$

Divide by 2. $y = \dfrac{96}{2} = 48$

As mentioned, in physics you will encounter many situations where the quantities are related in direct proportions. A simple example is the stretching of an ordinary helical spring. The spring has a certain length to begin with, and when it is suspended vertically with weights attached to the bottom, the length of the spring increases. If the amount of stretch is not too long, the amount of weight F (measured in pounds or newtons) and the amount of stretch x are directly proportional to each other (Figure 3.1). This is known as *Hooke's law*.

Thus, $F = k\,x$.
The constant k is called the *stiffness constant* or *spring constant* of the spring.

Figure 3.1

Example 3.4.2 (from mechanics)

A spring is suspended vertically from a fixed support (Figure 3.1). The stiffness constant of the spring is $k = 10.0$ N/m. When a weight is attached to the bottom of the spring, the spring stretches by 0.1 m. Determine the weight attached to the spring. The weight will be in newtons (N).

List data, check units, and convert the unit(s) if necessary. We have
$$k = 10.0 \text{ N/m}$$

and $x = 0.1 \text{ m}$

The units are correct; no conversion is necessary. The sketch of the problem is shown in Figure 3.1, where $x = 0.1$ m. The applied weight and the amount of stretch are related in direct proportions, as expressed by Hooke's law, $F = k\,x$. Apply Hooke's law to solve for the weight:

$$F = k\,x \ = (10.0 \text{ N/m}) \cdot (0.1 \text{ m}) = 1.0 \text{ N}$$

Example 3.4.3 (from thermal physics)

Charles's law is an important law governing the behavior of gases; it states that if the pressure of a gas is kept constant, its volume (V) is *directly proportional* to its absolute temperature (T) measured in kelvins (K). Use this relationship to determine the volume of 2 l of a gas kept in an open container when it heats up from 77 K to room temperature 300 K.

Initial volume $V_1 = 2\,\text{l}$
Initial temperature, $T_1 = 77$ K
Final temperature, $T_2 = 300$ K
Final volume, $V_2 = ?$

Since the container is open, pressure is always constant. Volume and temperature are related in direct proportions, since pressure is constant. Thus, the ratio V/T is constant, or

$$\frac{V_1}{T_1} = \frac{V_2}{T_2}$$

Multiply both sides by T_2 to solve for V_2. $V_2 = \dfrac{V_1}{T_1}\,T_2$

Substitute the values of V_1, T_1, and T_2. $V_2 = \left(\dfrac{2\,\text{l}}{77\,\text{K}}\right) 300 \text{ K} = 8\,\text{l}$

As temperature has increased, the final volume has increased, so the result makes sense.

Exercises

2. Hubble's law: According to Hubble's law, the speed of recession (v) of a galaxy is *directly proportional* to its distance (d) from the earth and is given by $v = Hd$, where the proportionality constant H is called Hubble's constant, which is 0.022 in the unit of m/(s · lightyear) (That is, the constant is 0.022 when speed is measured in the unit of *meter per seconds*, and the distance is measured in *lightyears*.) A galaxy is found to be moving away from the earth at a speed of 5.0 X 10^5 m/s. Determine the distance of the galaxy in lightyears.

3. A hand exerciser uses a coiled spring that follows Hooke's law, which states that the force needed to compress the spring is *directly proportional* to the amount of compression. If a force of 90 N is needed to compress the spring by 0.02 m, determine the force required to compress the spring by 0.047 m. (*Hint*: $F_1/x_1 = F_2/x_2$)

4. When the volume of a closed container is constant, the pressure P of gas inside the container is *directly proportional* to its temperature T in kelvins (K). An automobile tire is filled to a pressure of 2.1 X 10^5 Pascal (Pa) at a temperature of 283.0 K. After the car is driven for 1 hour, the temperature within the tire rises to 310.0 K. What is the final pressure of the gas?

5. The pressure of water inside a lake is directly proportional to its depth. If the pressure of water at a depth of 1.0 m is 9.8 X 10^3 N/m^2, what is the pressure of water at a depth of 18.0 m?

Graph of Direct Proportionality Relationship

   In a Cartesian graph a rectangular coordinate system is formed by two perpendicular lines. The horizontal line is called the x-axis and the vertical line is called the y-axis. The point where the axes cross is called the *origin*.

   When a quantity y has a direct proportionality with another quantity x, the graph of y versus x is always a straight line passing through the origin, as shown. The graph $y = kx$ is shown in Figure 3.2. In the graph, the change of the quantity x is labeled as Δx (which is often called "run") and the corresponding change in y is labeled as Δy (which is often called "rise").

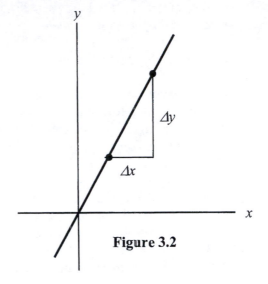

Here,

$$\Delta x = x_2 - x_1$$

$$\Delta y = y_2 - y_1$$

Figure 3.2

where (x_1, y_1) and (x_2, y_2) are coordinates of the two points on the line. The constant of proportionality between Δy and Δx is also k. Thus,

$$\Delta y = k \, \Delta x$$

The steepness of the line is measured by the ratio $\Delta y / \Delta x$ and is called the *slope* of the line. Thus,

$$\text{Slope} = \frac{y_2 - y_1}{x_2 - x_1} = \frac{\Delta y}{\Delta x} = k$$

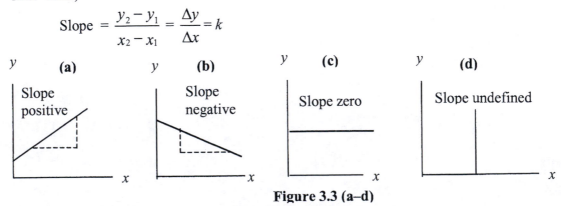

Figure 3.3 (a–d)

Therefore, in the direct proportionality between two quantities, the proportionality constant is the slope of the line representing the relationship between the quantities.

The slope of a line can be *positive, negative, zero* or *undefined*, as shown in Figure 3.3 (a–d).

Note:
Slope *positive* means *y increases as x increases.*
Slope *negative* means *y decreases as x increases.*
Slope *zero* means *y* does not change—that is, the line is *parallel to the x-axis.*
Slope *undefined* means *x* does not change—that is, the line is *parallel to the y-axis.*

3.5 Simple Inverse Proportion

In physics, you will encounter situations where one physical quantity increases and the other quantity decreases in such a way that their product always stays the same. In this case, they are said to be in *inverse proportion*. In inverse proportion, when one quantity approaches zero, the other quantity becomes extremely large, so that the product still remains the same. Mathematically, if y is in inverse proportion with x,

$$y \propto \frac{1}{x} \qquad \text{or} \qquad y = \frac{k}{x}, \text{ where } k \text{ is the proportionality constant}$$

Thus, when y is inversely proportional to x, $xy = k$.

That is,, when x changes from x_1 to x_2, y changes y_1 to y_2 so that $x_1 y_1 = x_2 y_2 = k$.

This type of behavior is illustrated Figure 3.4 (for an arbitrary choice of $k = 10$).

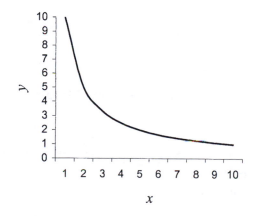

Figure 3.4

Example 3.5.1 (from thermal physics)

Boyle's law is an important law governing the behavior of ideal gases. According to the law, if the temperature of a gas is kept constant, its pressure, P, is *inversely proportional* to its volume, V. A cylindrical flask is fitted with an airtight piston. Contained within the flask is an ideal gas. Initially, the pressure applied by the piston is 12×10^4 Pa and the volume of the gas is 7.5×10^{-3} m^3. Assuming that the system is always at the temperature of 290 K, determine the volume of the gas when its pressure increases to 18×10^4 Pa. We have

$$P_1 = 12 \times 10^4 \text{ Pa} \qquad\qquad P_2 = 18 \times 10^4 \text{ Pa}$$
$$V_1 = 7.5 \times 10^{-3} \text{ m}^3 \qquad\qquad V_2 = ?$$

Following is a sketch of the problem:

Since the pressure P is inversely proportional to the volume V according to Boyle's law, the product PV remains constant.

That is,
$$P_1 V_1 = P_2 V_2$$

Divide by P_2 to solve for V_2.
$$V_2 = \frac{P_1 V_1}{P_2}$$

Substitute the values of P_1, V_1, and P_2.
$$= \frac{\left(12 \times 10^4 \text{ Pa}\right) \times \left(7.5 \times 10^{-3} \text{ m}^3\right)}{18 \times 10^4 \text{ Pa}}$$

$$= \frac{12 \times 7.5}{18} \times 10^{-3} \text{ m}^3$$

$$= 5 \times 10^{-3} \text{ m}^3$$

Since the pressure increased, the final volume had decreased, as expected.

Exercise

6. According to Kepler's second law of planetary motion, the speed v of a planet around the sun is *inversely proportional* to the distance r of the planet from the sun. If the speed of a planet is 5.0×10^4 m/s when it is at its shortest distance of

1.2 X 10^{15} m from the sun, find its speed when it is at its longest distance of 2.2 X 10^{15} m from the sun.

3.6 Inverse Square Proportion

Inverse square dependence with distance is found in different laws of nature. In physics, you will encounter situations where a physical quantity decreases uniformly in all directions with the square of the distance. For example, sound intensity (also light intensity), which is a measure of the sound power (or light power) distributed over unit area, decreases as the inverse square of the distance from the source of sound. The force of gravitation due to a spherical mass (such as a planet or a star) decreases as the inverse square of the distance from the center of the mass. The electrostatic force due to a point electric charge decreases as the square of the distance from the charge.

Mathematically, if y varies inversely with the square of x, then

$$y \propto \frac{1}{x^2} \qquad \text{or} \qquad y = \frac{k}{x^2}$$

where k is the proportionality constant.

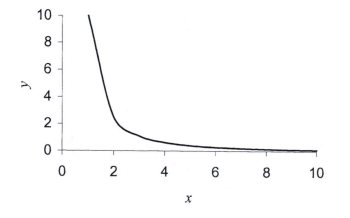

Figure 3.5

Example 3.6.1 (from sound)
A small source of sound emits sound equally in all directions. The intensity of sound is given by the equation $I = k/r^2$, where k is a constant and r is the distance from the source. If the sound intensity is 0.01 W/m² at a distance 1 m from the source, find the intensity at a distance of 10 m from the source.

Initial distance, $r_1 = 1$ m

Sound intensity, $I_1 = (0.01 \text{ W/m}^2)$

Final distance, $r_2 = 10$ m

Sound intensity, $I_2 = ?$

See the accompanying sketch of the problem.
    The intensity of sound is given by the equation $I = k/r^2$. Use this equation to solve the problem. However, in this problem, you do not need to know the value of k. From the inverse square relationship, we have

$$I_1 = \frac{k}{r_1^2} \qquad \text{and}$$

$$I_2 = \frac{k}{r_2^2}$$

Divide the preceding two equations to cancel k.

$$\frac{I_2}{I_1} = \frac{k/r_2^2}{k/r_1^2} = \frac{r_1^2}{r_2^2}$$

Multiply by I_1 to solve for I_2.

$$I_2 = I_1 \frac{r_1^2}{r_2^2}$$

Substitute the values of I_1, r_1, and r_2.

$$= (0.01 \text{ W/m}^2)\frac{(1\,\text{m})^2}{(10\,\text{m})^2}$$

$$= \frac{0.01}{100} \text{ W/m}^2 = 0.0001\,\text{W/m}^2$$

As the distance increases, the intensity decreases according to the equation $I = k/r^2$. That is, the value of the sound intensity at 10 m should be less that the value at 1 m. Thus, the result makes sense.

Example 3.6.2 (from electricity)

Two opposite kind electric charges (+ and −) attract while two same kind charges (either two + or two −) repel each other. The force of interaction between two electric charges is given by the Coulomb law: $F = kq_1q_2/r^2$, where q_1, q_2 are the charges, k is a constant, and r is the separation between them. A +7.0 μC and a −3.0 μC charge are palced 18.0 cm apart. Where can a third positive charge be placed so that it experiences no net force?

First charge, $q_1 = +7.0$ μC

Second charge, $q_2 = -3.0$ μC

Separation between the given charges, $r = 18.0$ cm

Let the third charge $= +q$ μC

The distance of the third charge from −3.0 μC $= x = ?$

The charge $+q$ will be repelled by the +7.0 μC charge and attracted by the −3.0 μC charge. Thus, the charge $+q$ will be placed along the line of the charges, but not between them. For the net force to be zero, the individual forces must be equal and opposite.

Thus, from the Coulomb law the magnitude

of the two forces
$$F = \frac{k(7.0\mu C)\, q}{(0.18 + x)^2} = \frac{k(3.0\mu C)\, q}{x^2}$$

Cancel same terms from both sides.
$$\frac{7}{(0.18 + x)^2} = \frac{3}{x^2}$$

Take the square root.
$$\frac{2.646}{0.18 + x} = \pm \frac{1.732}{x}$$

$$1.732\,(0.18 + x) = \pm 2.646\, x$$

$$0.3118 + 1.732x = \pm 2.646\, x$$

The solutions are
$$x = 0.34 \text{ m} \quad -0.071 \text{ m}$$

The negative result corresponds to the position between the charges (where the two forces act in the same direction and do not cancel). Thus, the third charge should be placed at 0.34 m; that is, 34 cm meter beyond the −3.0 C charge.

Exercises

7. A lamp emits light uniformly in all directions, and the intensity is given by an inverse square law, $I = k/r^2$. If the light intensity reaching an object at a distance of 0.20 m is 100.0 lux, determine the light intensity at 1.5 m from the lamp.

8. The force of gravity exerted by the earth on an object varies inversely as a square of the distance from the center of the earth. Determine the ratio of the force of gravity on an astronaut when he is on the surface of the earth and at a height of 10,000 km from the surface of the earth. The radius of the earth is 6.38×10^6 m.

9. Three particles of charge +4.0, +2.0, and +1.5 μC are placed on a line as shown. The charge +2.0 μC is 0.15 m to the right of the +4.0 μC charge and is 0.20 m to the left of the +1.5 μC charge. Determine the ratio of the forces given by +4.0 μC and +1.5 μC charges on the +2.0 μC charge.

Rules of Algebra

3.7   Power and Exponents

You will find that many equations in physics are nonlinear. For example, the relation between distance and time for a freely falling object (such as a falling apple from a tree) is nonlinear. In the expression x^m, x is called the *base* and m is called the *exponent*. In a nonlinear equation, one or more variables have exponents or powers.

$$x^m = x \cdot x \cdot x \cdot x \cdot x \cdot \ldots \cdot \text{(product of } m \text{ factors of } x)$$

Example 3.7.1 (from heat transfer)

Stefan's law: Energy radiation from hot bodies varies as the *fourth power* of its absolute temperature, $E = \sigma T^4$, where σ is a constant. Determine the factor by which the energy radiated from the sun will be reduced if its surface temperature changes from 6000 K to 5600 K.

For this problem, $T_1 = 6000$ K and $T_2 = 5600$ K.

Radiant energy will be E_1 and E_2 when the temperatures are T_1 and T_2, respectively. Use Stefan's law.

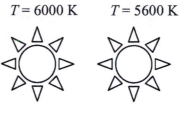

From Stefan's law,
$$E_1 = \sigma T_1^4$$
$$E_2 = \sigma T_2^4$$

Divide to cancel the constant σ.
$$E_1/E_2 = \sigma T_1^4/\sigma T_2^4 = T_1^4/T_2^4$$

Substitute the values of T_1 and T_2.
$$E_1/E_2 = (6000)^4/(5600)^4 = 1.3$$

Thus, the energy radiated from the sun will be reduced by the factor of 1.3. Since the temperature has decreased, the radiated energy has decreased, as expected from Stefan's law.

3.8   Rules for Exponents

Use the following rules for exponents, where m and n are natural numbers:

Product rule	$x^m \cdot x^n = x^{m+n}$
Quotient rule	$\dfrac{x^m}{x^n} = x^{m-n} \ (x \neq 0)$
First power rule	$(x^m)^n = x^{m \cdot n}$
More power rules	$(xy)^m = x^m \cdot y^m$
	$\left(\dfrac{x}{y}\right)^m = \dfrac{x^m}{y^m}$
Zero exponent rule	$x^0 = 1$ (where x is a nonzero real number; 0^0 is undefined)
Negative exponent rule	$x^{-m} = \dfrac{1}{x^m}$
Roots	$x^{1/m} = \sqrt[m]{x}$

The preceding rules can be verified easily. For example,

$$10^3 \times 10^2 = (10 \times 10 \times 10) \times (10 \times 10) = 10^5 = 10^{(3+2)}$$

This verifies the product rule. Similarly,

$$\frac{10^3}{10^2} = \frac{10 \times 10 \times 10}{10 \times 10} = 10 = 10^1 = 10^{(3-2)}$$

This verifies the quotient rule. Also,

$$(10^2)^3 = (10 \times 10)^3 = (100)^3 = 100 \times 100 \times 100 = 100\,00\,00 = 10^6 = 10^{3 \cdot 2}$$

This verifies the first power rule.

Note that the zero exponent rule follows from the quotient rule, since

$$1 = \frac{x^2}{x^2} = x^{2-2} = x^0$$

The negative exponent rule also follows from the zero exponent rule and the quotient rule because

$$\frac{1}{x^m} = \frac{x^0}{x^m} = x^{0-m} = x^{-m}$$

Example 3.8.1
Simplify each expression:

(i) $(-3x^4y^2)^2$

(ii) $\left(\dfrac{4x^2}{2y^3}\right)^2$

(iii) $\dfrac{\left(x^3y^4\right)^2}{\left(xy\right)^4}$

(iv) $\left(-\dfrac{x^2y^4}{xy^{-2}}\right)^{-3}$

(i) $(-3x^4y^2)^2 = (-3)^2(x^4)^2(y^2)^2 = 9x^{4 \cdot 2}y^{2 \cdot 2} = 9x^8y^4$

(ii) $\left(\dfrac{4x^2}{2y^3}\right)^2 = \dfrac{4^2x^{2 \cdot 2}}{2^2y^{3 \cdot 2}} = \dfrac{16x^4}{4y^6} = \dfrac{4x^4}{y^6}$

(iii) $\dfrac{\left(x^3 y^4\right)^2}{\left(xy\right)^4} = \dfrac{\left(x^3\right)^2 \left(y^4\right)^2}{x^4 y^4} = \dfrac{x^{3\cdot2} y^{4\cdot2}}{x^4 y^4} = \dfrac{x^6 y^8}{x^4 y^4} = x^{6-4} y^{8-4} = x^2 y^4$

(iv) $\left(-\dfrac{x^2 y^4}{x y^{-2}}\right)^{-3} = \left(-x^{2-1} y^{4-(-2)}\right)^{-3} = \left(-x^1 y^6\right)^{-3} = -x^{-3} y^{-18}$

Example 3.8.2 (from acoustics)

In a chromatic music scale, there are 12 equal music intervals, and the frequency of every thirteenth note is double that of the beginning note. That is, the frequency of each note increases by a factor of $2^{1/12}$ from the preceding note. If the frequency of the middle C on the piano is 261.63 Hz (Hz abbreviates hertz, the unit of frequency), what is the frequency of the next note in the music scale? (The frequency of a note tells how many cycles of vibration the source of sound completes in one second as it produces sound.)

Frequency of the middle C = 261.63 Hz.

Since the frequency of each note is increased by the factor $2^{1/12}$, to get the frequency of the next note, multiply that frequency by $2^{1/12}$.

Thus, frequency of the next note of middle C

$$= 261.63 \text{ Hz} \times 2^{1/12}$$

$$= 261.63 \text{ Hz} \times 2^{0.0833}$$

Use a calculator to find $2^{0.0833}$. $= 261.63 \times 1.0595 = 277.20 \text{ Hz}$

Exercises

Simplify each expression

10. $(x^2 y^4)^2 (x^3 z^2)^{-2} (z^3 x^2)^{-1}$

11. $\left(\dfrac{x^2 y^4}{x y^2}\right)^0$

12. $\dfrac{\left(x y^{-3} z\right)^3}{\left(x^{-2} y\right)^2}$

13. $\left(\dfrac{2a^2 b^4 c^{-2}}{8a^{-1} b^2 c^{-4}}\right)^{-3}$

3.9   Factoring a Polynomial and an Equation

An algebraic expression containing the sum of one or more terms that contain whole number exponents is called a *polynomial*. A polynomial that has only one term is called a *monomial*, whereas polynomials with two and three terms are called *binomials* and *trinomials*, respectively. For example, $2x + 4$ is a binomial and $5x^2 + 2x + 3$ is a trinomial. The *degree* of the monomial ax^n $(a \neq 0)$ is n. The degree of a monomial with more than one variable is the sum of the exponents of the variables. For example, the degree of the monomial $2x^2y^4$ is $(4 + 2)$ or 6.

An *equation* is an equivalent statement that indicates that the two quantities are equal. For example, $x - 2 = 4$ is an equation. Here, x is the variable of the equation. $x - 2$ is the left-hand side, and 4 is the right-hand side, of the equation. All the values of x that make the left-hand and right-hand sides equal are called solutions of the equation. An equation such as $a \cdot x + b = 0$ $(a \neq 0)$ that involves first-degree polynomials is called a *linear equation*. In general, an equation of the form $ax^2 + bx + c = 0$ (where a, b, and c are real numbers and $a \neq 0$) is called a *quadratic equation*.

Common terms can be *factored* out from a polynomial or an equation to simplify it. A trinomial of the form $ax^2 + bx + c$, where $a \neq 0$ and a, b, and c are all integers, will factor into two binomials with integer coefficients if the value of $b^2 - 4ac$ is a perfect square. By direct multiplication,

$$(x + p)(x + q) = x^2 + px + qx + pq = x^2 + (p + q)x + pq$$

Thus, the coefficient of the middle term is the sum of p and q and the last term is the product of p and q. Note that, in this case, $b^2 - 4ac = (p + q)^2 - 4pq = (p - q)^2$.

Example 3.9.1
Factor $x^2 - 2x - 63$.

Compare the given trinomial with $ax^2 + bx + c$.

In this case, $a = 1$, $b = -2$ and $c = -63$

$$b^2 - 4ac = (-2)^2 - 4 \cdot 1 \cdot (-63) = 4 + 252 = 256 = 16^2$$

Since $b^2 - 4ac$ is a perfect square, the given trinomial will factor into two binomials.

Now, try to write the coefficient (-2) as $(p + q)$ and the last term (-63) as pq.

Product of the Factors	Sum of the Factors
−63 (1)	−63 + 1 = −62
63 (−1)	63 − 1 = 62
7 (−9)	7 − 9 = −2
−7 (9)	−7 + 9 = 2

Obviously, the choice is $p = 7$ and $q = -9$, since, in this case, $p + q$ is -2 and pq is -63.

Thus,

$$x^2 - 2x - 6 = x^2 + 7x - 9x + 7(-9)$$
$$= x(x + 7) - 9(x + 7)$$
$$= (x + 7)(x - 9)$$

Remember the following formulas for factoring.

$$ax + ay + az = a(x + y + z) \qquad (x + y)^3 = x^3 + 3x^2y + 3xy^2 + y^3$$

$$x^2 - y^2 = (x + y)(x - y) \qquad (x - y)^3 = x^3 - 3x^2y + 3xy^2 - y^3$$

$$(x + y)^2 = x^2 + 2xy + y^2 \qquad x^3 + y^3 = (x + y)(x^2 - xy + y^2)$$

$$(x - y)^2 = x^2 - 2xy + y^2 \qquad x^3 - y^3 = (x - y)(x^2 + xy + y^2)$$

All of these formulas can be verified easily by direct multiplication.

Exercise

14. Factor the following trinomials:

    (a) $t^2 - 7t + 12$ (b) $2x^2 - 4x - 48$ (c) $96a + 6ab^2 + 48ab$

    (d) $2a^2 + 5a + 3$ (*Hint*: The first terms of the binomial factors are $2a$ and a.)

Solving Equations

3.10   Properties of an Equation

To solve an equation, remember the following *properties*.

If $x = y$ represents an equation and a is a real number, then

$$x \pm a = y \pm a$$

$$x \cdot a = y \cdot a$$

$$\frac{x}{a} = \frac{y}{a} \qquad \text{if } a \neq 0$$

$$x^n = y^n$$

$$\sqrt[n]{x} = \sqrt[n]{y}$$

Note: *Any mathematical operation that is applied to one side of an equation must also be applied to the other side.*

Important Note: If any side of an equation contains more than one term, the operation must be applied to the *entire side*. That is, if $c = a + b$,

$$3c = 3(a + b) \qquad \qquad (\text{not } 3a + b)$$

$$\frac{c}{2} = \frac{a}{2} + \frac{b}{2} = \frac{a+b}{2} \qquad (\text{not } \frac{a}{2} + b)$$

$$c^2 = (a + b)^2 \qquad \qquad (\text{not } a^2 + b^2)$$

$$\sqrt{c} = \sqrt{a + b} \qquad \qquad (\text{not } \sqrt{a} + \sqrt{b})$$

Example 3.10.1 (from electricity)
The equation $V = IR$ is called Ohm's law in electricity, where V, I, and R are voltage, current, and resistance, respectively. Solve the equation for the current I when $V = 120$ volts and $R = 10$ ohms. The unit of current is ampere.

For this problem, $V = 120$ volts $R = 10$ ohm

Use Ohm's law to solve the problem.

$$V = IR$$

Divide both sides by R to solve for I.

$$\frac{V}{R} = \frac{IR}{R} = I$$

Substitute the values of V and R.

$$I = \frac{V}{R} = \frac{120 \text{ volts}}{10 \text{ ohms}} = 12 \text{ amperes}$$

3.11   Linear Equations Involving One Variable

An equation of the form $ax + b = 0$ $(a \neq 0)$ is called a linear equation involving the unknown variable x. The basic operations of an equation (mentioned in the previous section) are sufficient to solve all linear equations. The objective is to select the proper operations to determine the unknown quantity x.

Example 3.11.1
Solve $3x + 2 = 5x - 1$.

$$3x + 2 = 5x - 1$$

Add -2 on both sides

$$3x + 2 - 2 = 5x - 1 - 2$$

$$3x = 5x - 3$$

Add $-5x$ on both sides so that right-hand side does not contain any term containing x.

$$3x - 5x = 5x - 3 - 5x$$

$$-2x = -3$$

Divide both sides by -2.

$$\frac{-2x}{-2} = \frac{-3}{-2}$$

$$x = \frac{3}{2}$$

To verify your answer, substitute $x = 3/2$ in the equation. The left-hand side is then 3 $(3/2) + 2 = 9/2 + 2 = 13/2$. The right-hand side is $5 (3/2) - 1 = 15/2 - 1 = 13/2$, as expected.

Exercises
Solve the following equations for x.

15. $4x + 8 = 0$ 16. $4x + 5 = 9x - 20$ 17. $\dfrac{3}{x-2} = \dfrac{4}{x-1}$

Example 3.11.2

If $F = kqQ/r^2$, solve for k.

Multiply by r^2.
$$Fr^2 = kqQ$$

Divide by qQ to cancel qQ from the right.
$$\frac{F r^2}{qQ} = \frac{k\cancel{qQ}}{\cancel{qQ}} = k$$

Therefore,
$$k = \frac{F r^2}{qQ}$$

Exercises

Solve the following equations for the unknown in terms of other quantities in the equations.

18. $F = GMm/r^2$; solve for G.   19. $(P + a/V^2)(V - b) = RT$; solve for P.

20. $L = L_0(1 + \alpha \Delta T)$; solve for α.   21. $1/d_0 + 1/d_i = 1/f$; solve for f.

3.12   Linear Equations Involving Two Variables

An equation of the form $ax + by + c = 0$ or of the form $y = mx + b$ is called a *linear equation* in x and y. Linear relationships between two physical variables are common. For example, when a moving object speeds up at a constant rate a, its velocity can be expressed by an equation $v = v_0 + at$, where v_0 is the initial velocity of the object and t represents time. When a solid is heated, its length, L, increases with temperature, T, and the relationship can be approximated by a linear equation $L = L_0 [1 + \alpha(T - T_0)]$, where L_0 and T_0 are the initial length and temperature, respectively, of the solid and α is a constant for the solid. There are many other situations when the relationship between two physical variables can be approximated as linear.

Example 3.12.1 (from kinematics)

When a moving object speeds up at a constant rate a (acceleration) starting from an initial velocity v_0, its final velocity (v) increases linearly with time (t) and is given by $v = v_0 + at$. An object starts with an initial velocity of 1.1 m/s and speeds up at a constant rate (called acceleration) of 2.0 m/s². What is the speed of the object after 10.0 s?

For this problem, $v_0 = 1.1$ m/s,

$a = 2.0$ m/s²,

$t = 10.0$ s, and

$v = ?$

The sketch of the problem is shown above. Use the relationship $v = v_0 + at$ to solve for v.

$$v = v_0 + at$$

Substitute the values of v_0, a, and t. $v = 1.1$ m/s $+ (2.0$ m/s²$) \times 10.0$ s

$= 21.1$ m/s

$= 21$ m/s (rounding to two significant digits)

Since the object is accelerating, the final velocity is more than the initial velocity. So the result makes sense.

Exercises

22. The ability of a material to resist the flow of electricity is called its *electrical resistance R*. The electrical resistance of a metal increases linearly with temperature T according to the equation $R = R_0[1 + \alpha_0(T - T_0)]$, where α_0 is a constant of the metal, called the *temperature coefficient of resistance*. Find the change in resistance, $R - R_0$, of platinum when the temperature change ($T - T_0$) is 100°C. α_0 for platinum is 0.003927/C°.

23. The increase in length, ΔL, of brass of length L for a temperature rise by ΔT can be approximated by the linear equation $\Delta L = \alpha L \ \Delta T$.   Find the temperature change of a piece of brass of length 2.0 m when its length increases by 1.0 mm. The constant α for brass is 19.0×10^{-6} (in the unit of /C°).

3.13   Solving Two Simultaneous Linear Equations

Two linear equations can be solved by the *substitution method*, the *addition method*, or the *graphical method*. The method of determinants (Cramer's rule) can also be used to solve two linear equations, but this method is more useful to solve several linear equations each containing more than two variables. The graphical method and the method of determinants are explained in appendices D and E, respectively.

Substitution Method

To solve two equations involving x and y by substitution

(i)   Solve one equation for one variable (say, x) in terms of the other variable (say, y).

(ii)   Substitute the resulting expression for the variable x into the other expression. Solve the new equation for y.

(iii)   Substitute the known value of y into one of the original equations to solve for x.

(iv)   Verify that x and y satisfy the original equations.

Example 3.13.1

Solve $\begin{cases} x - y = 2 \\ x + 2y = 8 \end{cases}$ by the method of substitution.

First equation:	$x - y = 2$
Add $+ y$ on both sides.	$x - y + y = 2 + y$
or	$x = 2 + y$
Second equation:	$x + 2y = 8$
Substitute $2 + y$ for x.	$(2 + y) + 2y = 8$

or	$2 + 3y = 8$
Subtract 2 from both sides.	$3y = 8 - 2 = 6$
Divide both sides by 3.	$y = \dfrac{6}{3} = 2$
Substitute $y = 2$ into the original equation, $x - y = 2$	
	$x - 2 = 2$
Add +2 on both sides.	$x = 4$
The solution is	$\begin{cases} x = 4 \\ y = 2 \end{cases}$

(To verify your answer, substitute $x = 4$ and $y = 2$ in the left-hand side of both of the equations, $x - y = 4 - 2 = 2$ and $x + 2y = 4 + 2 \cdot 2 = 4 + 4 = 8$. Thus, x and y satisfy both the equations.)

Addition Method

To solve two equations by the method of addition:

(i)   Write both equations in the form $ax + by = c$.

(ii)   Multiply (if necessary) one or both of the equations by nonzero quantities so that the coefficients of x (or y) in the equations become equal but have opposite signs.

(iii)   Add the resulting equations to eliminate x (or y), and solve for y (or x).

(iv)   Substitute the known value of y (or x) into one of the original equations to solve for x (or y).

(v)   Verify that x and y satisfy the original equations.

Example 3.13.2

Solve $\begin{cases} 2x - y = 6 \\ x + 2y = 8 \end{cases}$ by the addition method.

Multiply the second equation by –2.

$$2x - y = 6$$
$$-2x - 4y = -16$$

Add $-5y = -10$

Divide by –5. $y = \dfrac{-10}{-5} = 2$

Substitute $y = -2$ into the original equation.

$$2x - 2 = 6$$

Add 2 on both sides. $2x = 6 + 2 = 8$

Divide by 2. $x = 4$

The solution is $\begin{cases} x = 4 \\ y = 2 \end{cases}$

Exercises

Solve the following equations by the substitution method and the addition method.

24. $x + y = 3, x - y = 1$ 25. $\dfrac{x}{2} + \dfrac{y}{3} = 2, \ 2x - y = 1$

26. $5x - 4y = 1, 6y = 10x - 4$ 27. $x = 2(4 - y), 3x = 6(8 - 3y)$

28. $2(x + 3) + 7(2 - y) = 4, 2(x - 3) - 4(y - 3) = 2$

29. $\dfrac{x+3}{2} + \dfrac{y+2}{3} = 5, \ \dfrac{x-1}{2} + \dfrac{y-1}{3} = 2$

3.14   Solution of Quadratic Equations

An equation of the form $ax^2 + bx + c = 0$, where a, b, and c are real numbers and $a \neq 0$, is called a *quadratic equation*.

Sometimes you will encounter the simplest quadratic equation, such as $x^2 = a$. To solve this equation you need to take the square root of both sides:

$$\sqrt{x^2} = \sqrt{a}$$

$$x = \pm\sqrt{a}$$

The symbol $\pm$ means that $\sqrt{a}$ carries either a positive sign or a negative sign. This should be obvious since the square of both $+\sqrt{a}$ and $-\sqrt{a}$ give a. Usually, in a physical problem, the situation will determine which sign or signs are appropriate.

Important Note:

$$\sqrt{ab} = \sqrt{a} \times \sqrt{b}$$

But $\sqrt{a+b}$ is not equal to $\sqrt{a} + \sqrt{b}$.

Example 3.14.1
Solve $4x^2 - 2 = 2x^2 + 16$.

To solve this equation, rearrange the terms so that left-hand side contains only the term in x^2. To cancel 2 from the left-hand side, add 2 on both sides. Also, add $-2x^2$ on both sides to cancel $2x^2$ from the right. Thus, add $(2 - 2x^2)$ on both sides.

$$4x^2 - 2 = 2x^2 + 16$$

Add $(2 - 2x^2)$.

$$4x^2 - 2 + (2 - 2x^2) = (2 - 2x^2) + 2x^2 + 16$$

or

$$4x^2 - 2x^2 = 2 + 16$$

$$2x^2 = 18$$

Divide both sides by 2.

$$x^2 = 9$$

Take the square root.

$$x = \pm\sqrt{9} = \pm 3$$

Substitute $x = \pm 3$ in the equation and verify that the answers are correct.

Example 3.14.2 (from kinematics)

When an object is dropped from a certain height the vertical height of fall is given by $y = (-4.98 \text{ m/s}^2) \, t^2$, assuming that the upward direction is positive and that t is measured in seconds. If a ball is dropped from a height of 100.0 m, when will it reach the ground?

List the data with the proper sign. Since the ball falls downward and the upward direction is assumed to be positive, in this case

$$y = -100.0 \text{ m}$$

Use the equation $y = -(4.9 \text{ m/s}^2) \, t^2$ to solve the problem.

$$y = -(4.9 \text{ m/s}^2) \, t^2$$

Substitute the value of y. $-100.0 \text{ m} = -(4.9 \text{ m/s}^2) \, t^2$

Divide by -9.8. $t^2 = \dfrac{-100.0 \text{ m}}{-4.9 \text{ m/s}^2}$

$$= 20.4 \text{ s}^2.$$

Take the square root. $t = \pm\sqrt{20.4} \text{ s} = \pm 4.5 \text{ s}$

Positive

100 m

y

Ground

The time of fall must be positive, so the only acceptable answer is $t = 3.2$ s.

General Solution of a Quadratic Equation

A *quadratic equation* of the form $ax^2 + bx + c = 0$ has the general solution

$$x = \frac{-b \pm \sqrt{b^2 - 4ac}}{2a}$$

The two roots, x_1 and x_2, are given by

$$x_1 = \frac{-b + \sqrt{b^2 - 4ac}}{2a}, \qquad x_2 = \frac{-b - \sqrt{b^2 - 4ac}}{2a}$$

If $b^2 = 4ac,$ the roots are equal and real.

If $b^2 > 4ac,$ the roots are unequal and real.

If $b^2 < 4ac,$ the roots are complex (containing real and imaginary parts).

Although a quadratic equation may have two real unequal roots, often only one answer is physically acceptable.

Example 3.14.3

Solve the equation $2x^2 + 6x - 18 = 0$.

Compare this equation with $ax^2 + bx + c = 0$. In this case, $a = 2$, $b = 6$, and $c = -18$. Therefore,

$$x = \frac{-b \pm \sqrt{b^2 - 4ac}}{2a}$$

Substitute the values.

$$= \frac{-6 \pm \sqrt{6^2 - 4 \cdot 2(-18)}}{2 \cdot 2} = \frac{-6 \pm \sqrt{36 + 144}}{4}$$

$$= \frac{-6 \pm \sqrt{180}}{4} = \frac{-6 \pm 6\sqrt{5}}{4} = \frac{-3 \pm 3\sqrt{5}}{2}$$

Thus, the roots are

$$x_1 = \frac{-3 + 3\sqrt{5}}{2}, \qquad x_2 = \frac{-3 - 3\sqrt{5}}{2}$$

Example 3.14.4 (from kinematics)

The following equation describes how the position y of an object undergoing a uniform acceleration a depends on time t: $(y - y_0) = v_0 t + \frac{1}{2} at^2$. Here, v_0 is the initial speed of the object and y_0 is its initial position.

A ball is thrown downward from the top of a 20.0 m high building with an initial speed of 2.0 m/s. The acceleration of such a freely falling object near the earth is known to be 9.8 m/s². Determine the time when the ball will reach the ground.

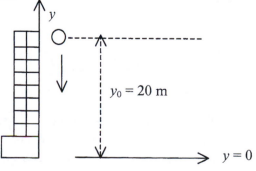

Consider the y-axis to be pointed upward.

Then $\qquad y_0 = +20.0$ m,

$\qquad\qquad v_0 = -2.0$ m/s,

$\qquad\qquad a = -9.8$ m/s^2, and

$\qquad\qquad y = 0$ (since the ball reaches the ground)

Use the equation $(y - y_0) = v_0 t + \dfrac{1}{2} a t^2$ to solve the problem. Rearranging the equation,

we get $\qquad\qquad\qquad\qquad\qquad \dfrac{1}{2} a t^2 + v_0 t - (y - y_0) = 0$

Substitute the values of y_0, v_0, y, and a $\qquad \dfrac{1}{2}(-9.8)t^2 + (-2.0)t - (0 - 20.0) = 0$

or $\qquad\qquad\qquad\qquad\qquad\qquad 4.9\ t^2 + 2t - 20 = 0$

This is a quadratic equation in t. Compare it with $ax^2 + bx + c = 0$. In this case, $a = 4.9$, $b = 2$, and $c = -20$.

The solutions are $\qquad\qquad\qquad t = \dfrac{-2 \pm \sqrt{2^2 - 4 \times 4.9\,(-20)}}{2 \times 4.9}$

$$= \dfrac{-2 \pm \sqrt{4 + 392}}{9.8} = \dfrac{-2 \pm 19.9}{9.8}$$

$$= 1.8 \text{ s or } -2.2 \text{ s}$$

The answer $t = -2.2$ s is not an acceptable solution because the ball must have fallen after it was dropped (i.e., the time cannot be negative). Thus, the ball reaches the ground in 1.8 seconds.

Exercises

30. Solve the following quadratic equations
    (a) $-200 = -16t^2$ $\qquad\qquad$ (b) $\quad 3t^2 - 2t - 3 = 0$ $\qquad\qquad$ (c) $y^2 + y + 1 = 0$

31. An object is dropped from a height of 40 m. When will the object reach the ground? Assume that its vertical position y is related to its time of fall t by $y = v_0 t + \frac{1}{2} a t^2$. The acceleration a is directed downward and is 9.8 m/s^2 in magnitude.

3.15   Inequalities

An *inequality* is a mathematical statement of the fact that two quantities are not, or may not be, equal. For example, $x + 2 \geq 7$ is an inequality. Any value of x that satisfies the inequality is a solution of the inequality. Usually, an inequality has many solutions.

Assume that x, y, and a are real numbers, and use the following addition, subtraction, multiplication, and division properties of an inequality:

Addition and subtraction property   If $x < y$, then $x \pm a < y \pm a$.

Example: $2 < 4$, so if we add 3 on both sides, we get $2 + 3 < 4 + 3$, or $5 < 8$, as expected.

Multiplication property   If $x < y$ and $a > 0$, then $xa < ya$.
                                    If $x < y$ and $a < 0$, then $xa > ya$.

Example: $2 < 4$, so if we multiply by 3, we get, $2 \cdot 3 < 4 \cdot 3$, or $6 < 12$, as expected.

                                    $2 < 4$, so if we multiply by -3, we get, $2 \cdot (-3) > 4 \cdot (-3)$, or $-6 > -12$, as expected.

Division property   If $x < y$ and $a > 0$, then $\dfrac{x}{a} < \dfrac{y}{a}$.

                                    If $x < y$ and $a < 0$, then $\dfrac{x}{a} > \dfrac{y}{a}$.

Example: $2 < 4$, divide by 2, we get $2/4 < 4/2$, that is, $\frac{1}{2} < 2$, as expected.
                                    Also, $2 < 4$, divide by -2, we get $2/-2 > 4/-2$, that is, $-1 > -2$, as expected.

Similar statements hold for $>$, $\geq$, and $\leq$.

Note: If both sides of an inequality are multiplied or divided by a *negative* number, the direction of the inequality is *reversed*.

Example 3.15.1

Solve $3 - 2x \leq 11$.

$$3 - 2x \leq 11$$

Add –3 on both sides. $-2x \leq 11 - 3$

or $-2x \leq 8$

Divide by –2. $x \geq -\dfrac{8}{2}$ (Note the change of the

                                                 direction of the inequality.)

or $x \geq -4$

Check answer by placing a number, such as $x = -3$, in the original inequality to verify the solution.

Example 3.15.2 (from thermal physics)

Room temperature is expressed in the Fahrenheit scale (F) in the United States and in the Celsius scale (C) in Canada. The temperature of a room in the United States is maintained between 68°F and 71°F. What would be the corresponding range in the Celsius scale? The temperatures in the two scales are related by the equation $F = \dfrac{9}{5} C + 32.$

We have

$$68 \leq F \leq 71$$

Substitute $F = \dfrac{9}{5} C + 32.$ $68 \leq \dfrac{9}{5} C + 32 \leq 71$

Add –32. $68 - 32 \leq \dfrac{9}{5} C \leq 71 - 32$

Multiply by 5/9. $36 \cdot \dfrac{5}{9} \leq C \leq 39 \cdot \dfrac{5}{9}$

                                         $20 \leq C \leq 21.7$

Thus, the temperature of the room must be maintained between 20°C and 21.7°C.

Exercises

32. Solve
    (a) $2x + 7 > x - 2$ (b) $x - 4(x - 1) > 11$ (c) $-4 < 3(x - 1) \leq 10$

33. In the first two physics tests, a student has test scores of 88% and 87%. What must he score on the third test to earn 90% or more cumulatively?

Logarithms and the Exponential Function

3.16   Logarithms

If $y = b^x$, then x is called the *logarithm* of y relative to b. This is written as $\log_b y = x$. Thus, the *logarithm* of a positive number, y, relative to a positive base, b (different from 1), is defined as the exponent or the power to which the base b is raised to produce the number y.

The following rules apply to logarithms:

$\log_b (mn) = \log_b m + \log_b n$ (rule for addition)

$\log_b \left(\dfrac{m}{n} \right) = \log_b m - \log_b n$ (rule for subtraction)

$\log_b (m^n) = n \log_b m$ (rule for powers)

$(\log_a m)(\log_m n) = \log_a n$ (rule for switching base)

The preceding rules can easily be verified by choosing any base b (different from 1) and any positive numbers m and n. Show that the left-hand side and right-hand sides are equal.

For example,   let us choose $b = 10$, $m = 3$ and $n = 2$. Then

$$\log_b (mn) = \log_{10}(3 \cdot 2) = \log_{10} 6 = 0.77815 \text{ (result from a calculator)}$$

Also, $\log_b m + \log_b n = \log_{10} 3 + \log_{10} 2 = 0.47712 + 0.30103 = 0.77815$

Therefore, $\log_{10} (3 \cdot 2) = \log_{10} 3 + \log_{10} 2$

This verifies the addition rule: $\log_b (mn) = \log_b m + \log_b n$.

Similarly, other rules can easily be verified.

Example 3.16.1

Write the expression $\log_b \dfrac{x \, y^4}{z^2}$ in terms of logarithms of x, y, and z.

Use the rule for subtraction.

$$\log_b \frac{x\,y^4}{z^2} = \log_b(x\,y^4) - \log_b z^2$$

Use the rule for addition.

$$= (\log_b x + \log_b y^4) - \log_b z^2$$

Use the rule for powers.

$$= \log_b x + 4\log_b y - 2\log_b z$$

Example 3.16.2

Simplify the expression $\log_b\left(\dfrac{x}{y} + x\right) - \log_b\left(\dfrac{z}{y} + z\right)$.

We have

$$\log_b\left(\frac{x}{y} + x\right) - \log_b\left(\frac{z}{y} + z\right)$$

Use the rule for subtraction.

$$= \log_b \frac{\left(\dfrac{x}{y} + x\right)}{\left(\dfrac{z}{y} + z\right)}$$

Multiply the numerator and denominator by y.

$$= \log_b \frac{(x + xy)}{(z + zy)}$$

Factor out x and z.

$$= \log_b \frac{x(1+y)}{z(1+y)} = \log_b \frac{x}{z}$$

Exercises

34. Use the properties of logarithms and write each expression in terms of logarithms of x, y, and z.

(a) $\log_b(x^3 y^2 z^{10})$ (b) $\log_b \sqrt[3]{\dfrac{x^2 y^3}{z^2}}$ (c) $\log_b \sqrt{x^2 y^4}$

35. Simplify each expression.

(a) $4\log_b x + \dfrac{1}{2}\log_b y - 2\log_b(x+y)$ (b) $\log_b(xy + x^2) + \log_b(xz + yz) - \log_b z$

Common Logarithms

For many logarithmic computations, it is common to use $b = 10$ as the base. In this case, the logarithm is called a *common logarithm*. Usually, log y means $\log_{10} y$. Since logarithm with respect to base10 appears very often, it is useful to remember that

log 1 = 0	because	$10^0 = 1$,
log 10 = 1	because	$10^1 = 10$,
log 100 = 2	because	$10^2 = 100$,
log 1000 = 3	because	$10^3 = 1000$,
log (0.1) = −1	because	$10^{-1} = 0.1$,
log (0.01) = −2	because	$10^{-2} = 0.01$,
log (0.001) = −3	because	$10^{-32} = 0.001$

Before calculators were invented, extensive tables provided logarithms of numbers. It is now very easy to calculate the logarithm of a number using a calculator. You should read the manual of your calculator to calculate logarithms and determine its full abilities. Usually, you type the number, press the LOG key, and then press the key that has sign =, the equals sign. For example, to find log 32.58, enter 32.58 and press LOG. To use a graphing calculator, enter LOG 32.58 and then ENTER. To determine a number whose logarithm is known, usually you enter the known number, press the 10^x key, and then press the = key. The plot of $y = \log_{10}x$ is shown in Figure 3.6.

The base 10 logarithm is used often in physics, such as to define the sound intensity level in acoustics and the voltage gain in electronics. The logarithmic unit of sound intensity level or voltage gain is called the decibel (dB). In geophysics, the base 10 logarithm is used to measure the intensity of earthquakes on the Richter scale. The base 10 logarithm is also used in chemistry to determine the hydrogen ion concentration of a solution. The next several examples and exercises will illustrate the use of the base 10 logarithm in these cases.

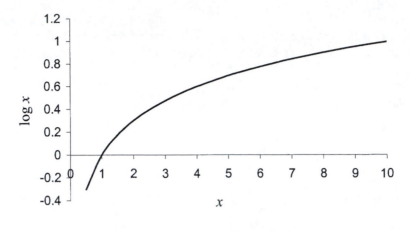

Figure 3.6

Example 3.16.3 (from acoustics)

The base 10 logarithm is used in physics to determine the sound intensity level in decibels (dB), which is a measure of the loudness of sound. The sound intensity level β is defined as $\beta = 10 \log \dfrac{I}{I_0}$, where I is the intensity of sound in watts/square meter (W/m^2) and $I_0 = 10^{-12}$ W/m^2, called the threshold intensity (minimum sound intensity) of hearing. Determine the sound intensity level at a point where the sound intensity is $I = 10^{-4}$ W/m^2.

For this problem, $I = 10^{-4}$ W/m^2

Also, $I_0 = 10^{-12}$ W/m^2

$$\beta = ?$$

Following is a sketch of the problem:

$\vartriangleleft$-------------------------------------● $I = 10^{-4}$ W/m^2

Use the expression for sound level, $\beta = 10 \log \dfrac{I}{I_0}$, to solve the problem. The unit of β is dB.

Therefore, $\beta = 10 \log \dfrac{I}{I_0}$

Substitute the values of I and I_0.

$$= 10 \log \frac{10^{-4}\ \mathrm{W/m^2}}{10^{-12}\ \mathrm{W/m^2}} = 10 \log (10^8)\ \mathrm{dB}$$

Use the rule for powers.

$$= 10 \times 8 \log 10\ \mathrm{dB}\ (\text{since } \log (m^n) = n \log m)$$

$$= 80 \log 10\ \mathrm{dB}$$

$$= 80 \times 1\ \mathrm{dB} \qquad (\text{since } \log 10 = 1)$$

$$= 80\ \mathrm{dB}$$

Example 3.16.4 (from acoustics)

The sound intensity level at a point is 91.0 dB. Determine the sound intensity.

For this problem,

$$\beta = 91.0\ \mathrm{dB}$$

$$I_0 = 10^{-12}\ \mathrm{W/m^2}$$

$$I = ?$$

Therefore,

$$\beta = 10 \log \frac{I}{I_0}$$

Substitute the values of β and I_0.

$$91.0\ \mathrm{dB} = 10 \log \frac{I}{10^{-12}\ \mathrm{W/m^2}}$$

Divide by 10.

$$\log \frac{I}{10^{-12}\ \mathrm{W/m^2}} = \frac{91}{10} = 9.1$$

$$\frac{I}{10^{-12}\ \mathrm{W/m^2}} = \mathrm{INV}\ \log\ (9.1)\ (\text{which}$$

$$\text{is } 10^{9.1})$$

Take the inverse logarithm.

$$= 1.26 \times 10^9\ \mathrm{W/m^2}$$

Multiply by 10^{-12}.

$$I = 10^{-12} \times (1.26 \times 10^9)\ \mathrm{W/m^2}$$

$$= 1.3 \times 10^{-3}\ \mathrm{W/m^2}$$

Example 3.16.5 (from electronics)

The base 10 logarithm is also used in electronics to determine the voltage gain of an electronic device, such as an electronic amplifier. The voltage gain is expressed in decibels. If V_0 is the output voltage and V_i is the input voltage of an amplifier, its decibel voltage gain is given by $20 \log \dfrac{V_0}{V_i}$. Determine the decibel voltage gain of an amplifier that provides an output voltage of 56 V when the input voltage is 0.11 V.

For this problem, $V_0 = 56$ V,

$\qquad\qquad\qquad\quad V_1 = 0.11$ V

Use the expression voltage gain $= 20 \log \dfrac{V_0}{V_i}$ to determine the voltage gain in decibel

(dB).

$$\text{Voltage gain } = 20 \log \frac{V_0}{V_i} = 20 \log \frac{56\,\text{V}}{0.11\text{V}} = 20 \log (509.1)$$

$$= 20 \times 2.7 = 54 \text{ dB}$$

Example 3.16.6 (from chemistry)

To express the acidity of a solution, the common (base 10) logarithm is used. The concentration is expressed by pH, which is defined as pH $= -\log [H^+]$, where H^+ is the hydrogen ion concentration of the solution in gram-ions per liter. Determine the hydrogen ion concentration of a solution if its pH is 8.2.

We have   pH $= 8.2$.

Use the equation pH $= -\log [H^+]$ to calculate the hydrogen ion concentration [H+].

$$\text{pH} = -\log [H^+]$$

Substitute the value of pH. $8.2 = -\log [H^+]$

or $\log [H^+] = -8.2$

$[H^+] = \text{Inv} \log (-8.2)$ (This is $10^{-8.2}$.)

Take the inverse logarithm. $[H^+] = 6.3 \times 10^{-9}$ gram-ions per liter.

$\qquad\qquad\qquad\qquad\qquad\qquad\qquad\qquad$ (Use a calculator.)

Exercises

36. Determine the sound intensity at a point where the sound intensity level is 50 dB.

37. The threshold of pain (the maximum sound that you can hear without causing any problem) is 120 dB. Determine the corresponding sound intensity, whose unit is watts per square meter (W/m^2).

38. Determine the voltage gain of an electronic amplifier whose input and output voltages are 2.0 mV and 105.0 mV.

39. The pH of pure water is 7. Determine the hydrogen ion concentration of pure water in gram-ions per liter.

40. Find the pH of a solution if its hydrogen ion concentration is 4.6×10^{-7} gram-ions per liter.

41. In seismology the intensity of an earthquake on the Richter scale R is expressed as $R = \log \dfrac{A}{T}$, where A is the amplitude (measured in micrometers) and T is the period (the time of vibration of the earth's surface, measured in seconds.) Determine the measure of an earthquake on the Richter scale with amplitude $A = 10{,}000$ micrometers and period $T = 0.15$ s.

Natural Logarithms

In many applications another base, called the natural base $e = 2.718...$, is used. Mathematically, the number e is the value of the limit, $\lim\limits_{n \to \infty} (1 + 1/n)^n$, which is equivalent to $\lim\limits_{z \to 0} (1 + z)^{1/z}$, where $z = 1/n$. The preceding limits mean that if you calculate the value of $(1 + 1/n)^n$ for different values of n [or the value of $(1 + z)^{1/z}$ for values of z], you will find that as n becomes larger and larger (or z becomes closer and closer to zero), the value of $(1 + 1/n)^n$ [or the value of $(1 + z)^{1/z}$] will become 2.718... For example, when

$$z = 1, \qquad (1 + z)^{1/z} = 2.0,$$
$$z = 0.5 \qquad (1 + z)^{1/z} = 2.25,$$
$$z = 0.25 \qquad (1 + z)^{1/z} = 2.4414,$$
$$z = 0.1 \qquad (1 + z)^{1/z} = 2.5937,$$
$$z = 0.01 \qquad (1 + z)^{1/z} = 2.7048,$$
$$z = 0.001 \qquad (1 + z)^{1/z} = 2.7169,$$
$$z = 0.0001 \qquad (1 + z)^{1/z} = 2.7181$$

Clearly, the expression $(1 + z)^{1/z}$ approaches the value 2.718... as z becomes closer and closer to zero.

Example 3.16.7

If A_0 dollars are put into an account that pays an annual interest rate of r, compounded N times in one year, it can be shown that the amount of money after t years becomes $A(t) = A_0 (1 + r/N)^{Nt}$. If the interest is compounded continuously, show that $A(t) = A_0 e^{rt}$. Determine the balance in the account after five years at the 4% rate of interest if the initial deposit A_0 is $100.

Since the interest is compounded continuously, it follows that $N = \infty$, and the balance after t years is

$$A(t) \quad = \lim_{N \to \infty} A_0 (1 + r/N)^{Nt}$$

$$= A_0 \lim_{n \to \infty} [(1 + r/N)^{N/r}]^{rt}$$

Substitute $n = N/r$. $= A_0 [\lim_{n \to \infty} (1 + 1/n)^n]^{rt}$ [as $N \to \infty$, $n \to \infty$].

Substitute e for the limit. $= A_0 e^{rt}$

For the next part of the problem substitute $A_0 = \$100$, $r = 4\% = 0.04$, and $t = 5$ years

Therefore, $A(t) \quad = A_0 e^{rt} = \$100 \ e^{0.04 \times 5} = \$122.14$

Likewise, the common logarithm is defined when $y = 10^x$, and the logarithm with respect to the base e, known as *natural logarithm* and denoted by $\ln x$, is defined when $y = e^x$. The exponential function e^x and natural logarithm occur in several natural phenomena where the rate of change of some quantity depends on the quantity. A plot of $y = \ln (x)$ is shown in Figure 3.7.

Note that $e^{\ln x} = x$. This can be verified easily by taking the natural logarithm of $e^{\ln x}$.

$$\ln (e^{\ln x})$$

Use the power rule. $= \ln x \ln e$

By definition, $\ln e = 1$. $= \ln x \cdot 1 = \ln x$

Therefore, $e^{\ln x} = x$. Because of their inverse relationship, e^x and $\ln x$ are called *inverse functions*.

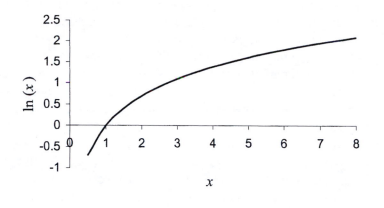

Figure 3.7

3.17   Exponential Functions

Exponential functions play a special role in mathematics and physics. The inverse of the natural logarithm, ln x is the exponential function, $y = e^x$, also written as exp (x). Similarly, the negative exponent exponential function is $y = e^{-x}$, also written as exp ($-x$). The plots of these exponential functions are shown in Figures 3.8 and 3.9, respectively, for $x = -1$ to 2.

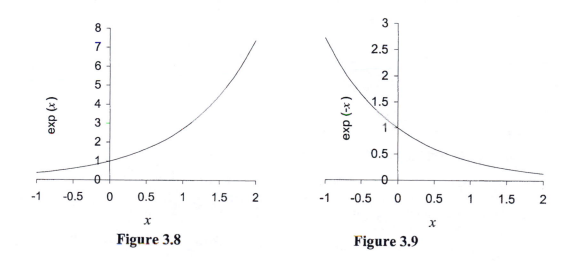

Figure 3.8   **Figure 3.9**

When a quantity increases like the positive exponent exponential function, it is called *exponential growth*. Similarly, when a quantity decreases like the negative exponent exponential function, it is called *exponential decay*.

From the plot of the function $y = \exp(x)$ (that is, e^x), it is easily seen that the function is a curve that has small values when its slope is small and large values when the slope is large. Thus, the rate of change of the curve $y = e^x$ is e^x. Similarly, the rate of decrease of the curve $y = e^{-x}$ is e^{-x}. This property is special in physics, since it occurs naturally, such as in the radioactive decay of an unstable nucleus of an atom. For radioactive decay, the rate of decay of the number of nucleus, N, with time t is proportional to N. This can be expressed as $\Delta N / \Delta t \propto N$. Thus, the number decays exponentially. Note that when there is sufficient food and space, populations of living organisms also tend to increase exponentially.

Note: The exponential function e^x and e^{-x} and natural logarithm occur in several natural phenomena in which the *rate of change (increase or decrease) of some quantity depends on the quantity.*

Example 3.17.1 (from nuclear physics)

Radon is radioactive and it decays. The number N of radon atoms remaining after time t (in days) is related to the original number N_0 of radon atoms by the expression $N = N_0\, e^{-(0.181/\text{day})t}$. If $N_0 = 2 \times 10^7$, find the time after which 2×10^5 radon atoms will remain.

For this problem, $N = 2 \times 10^5$

$N_0 = 2 \times 10^7$

Use the equation $N = N_0\, e^{-(0.181/\text{day})t}$ to determine the time t. Note that the power of e must be dimensionless. Thus, the unit of the time t is "day." We have

$$N = N_0\, e^{-(0.181/\text{day})t}$$

Substitute the values of N and N_0. $2 \times 10^5 = 2 \times 10^7\, e^{-(0.181/\text{day})t}$

Divide by 2×10^7. $\dfrac{2 \times 10^5}{2 \times 10^7} = e^{-(0.181/\text{day})t}$

or $10^{-2} = e^{-(0.181/\text{day})t}$

Change the equation into logarithmic form.	$\ln(10^{-2}) = -(0.181/\text{day})t$
Use the rule for powers, $\ln(m^n) = n \ln m$.	$-2 \ln 10 = -(0.181/\text{day})t$
Divide by –0.181.	$t = \dfrac{2 \ln 10}{0.181/\text{day}}$
	$= \dfrac{2 \times 2.303}{0.181/\text{day}} = 25.4 \text{ days}$

Example 3.17.2 (from electricity)

An electric circuit contining a resistance R and a capacitance C is called RC circuit, which is common such as used to change the speed of windshield wiper of a car. The growth of electric charge q with time t in a RC circuit containg a battery of emf ε is given by $q = C\varepsilon(1 - e^{-t/RC})$. Determine the time taken by a charge to reach 99 % of its final value when $R = 1$ KΩ (kiloohms) abd $C = 10$ μF (microfarads).

$$q = C\varepsilon(1 - e^{-t/RC})$$

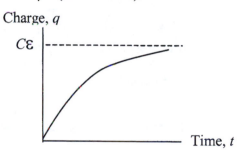
Charge, q

When $t = \infty$, $e^{-t/RC} = 0$. Thus, from the preceding equation, $q = C\varepsilon$ when $t = \infty$.

That is, $C\varepsilon$ represents the final value of the charge, q.

In the problem, then, $q = 99\% \ C\varepsilon = 0.99C\varepsilon$
$$t = ?$$

$R = 1$ K$\Omega = 1000$ Ω
$C = 10$ μF $= 10 \times 10^{-6}$ F

Use the equation	$q = C\varepsilon(1 - e^{-t/RC})$
Substitute $q = 0.99\ C\varepsilon$.	$0.99\ C\varepsilon = C\varepsilon(1 - e^{-t/RC})$
Cancel $C\varepsilon$ from both sides.	$0.99 = 1 - e^{-t/RC}$
	$e^{-t/RC} = 0.01$

Write the equation in logarithm form. $-t/RC = \ln(0.01)$

$t = -RC \ln(0.01)$

Substitute the values of R and C. $t = -(1000\ \Omega)(10 \times 10^{-6}\ F) \ln(0.01)$

$= 46 \times 10^{-3}\ s. = 46\ ms$

Exercises

42. The number of atoms, N, that remain of an unstable radioactive atom is given by the expression $N = N_0\ e^{-\lambda t}$. If the initial number of atoms is $N_0 = 2.4 \times 10^{15}$ and after 20 days the number of remaining atoms is $N = 1.2 \times 10^{14}$, determine the constant λ. The constant λ, called the decay constant, has unit day^{-1}.

43. The growth of bacteria in a laboratory culture increases according to the expression $N = N_0\ e^{(0.37/s)\ t}$. If the initial population is 500, determine the time when the population will be 5000.

44. The charge q of a capacitor decreases exponentially through a RC circuit from its initial value Q_0 according to the equation $q = q_0\ e^{-t/RC}$. If $R = 1000\ \Omega$ and $C = 1.0 \times 10^{-6}\ F$, determine the time the charge on the capacitor reduces to 37 % of its initial value.

The Binomial Theorem and Series Expansion

3.18   The Binomial Theorem

Consider the following *binomial expansions*, which you can verify by direct multiplication:

$$(a + b)^2 = (a + b)(a + b) = a^2 + ab + ba + b^2 = a^2 + 2ab + b^2$$

$$(a + b)^0 = 1$$
$$(a + b)^1 = a + b$$
$$(a + b)^2 = a^2 + 2ab + b^2$$
$$(a + b)^3 = a^3 + 3a^2b + 3ab^2 + b^3$$
$$(a + b)^4 = a^4 + 4a^3b + 6a^2b^2 + 4ab^3 + b^4$$

In general, according to the *binomial theorem*, for any positive integer n,

$$(a+b)^n = a^n + \frac{n!}{1!(n-1)!}\,a^{n-1}b + \frac{n!}{2!(n-2)!}\,a^{n-2}b^2 + \frac{n!}{3!(n-3)!}\,a^{n-3}b^3 + \ldots$$

$$+ \frac{n!}{p!(n-p)!}\,a^{n-p}b^p + \ldots + b^n$$

$$= \sum_{p=0}^{n} \frac{n!}{(n-p)!\,p!}\,a^{n-p}\,b^p$$

In the preceding expressions, the symbol ! is called factorial. $n!$ is defined as $n! = n(n-1)(n-2)\ldots\ldots 3 \cdot 2 \cdot 1$.

For example, $6! = 6 \cdot 5 \cdot 4 \cdot 3 \cdot 2 \cdot 1 = 720$.

Also, $n! = n(n-1)(n-2)\ldots\ldots 3 \cdot 2 \cdot 1 = n[(n-1)(n-2)\ldots\ldots 3 \cdot 2 \cdot 1] = n(n-1)!$.

Example 3.18.1

Expand $(3x + 2y)^4$, using the binomial theorem.

$$(3x+2y)^4 \quad = (3x)^4 + \frac{4!}{1!(4-1)!}\,(3x)^{4-1}(2y) + \frac{4!}{2!(4-2)!}\,(3x)^{4-2}(2y)^2$$

$$+ \frac{4!}{3!(4-3)!}\,(3x)^{4-3}(2y)^3 + (2y)^4$$

$$= (3x)^4 + 4\,(3x)^3(2y) + 6\,(3x)^2(2y)^2 + 4\,(3x)\,(2y)^3 + (2y)^4$$

$$= 81x^4 + 216x^3y + 216x^2y^2 + 96xy^3 + 16y^4$$

Exercise

45. Use the binomial theorem to expand these expressions:

(i) $(a+b)^5$

(ii) $\left(\dfrac{a}{3} - \dfrac{b}{2}\right)^4$

3.19   Power Series Expansions and Approximations

The right-hand side of the expressions that will follow usually contains an infinite number of terms. A polynomial with an infinite number of terms is called a *power series*; hence, the following expansions are called *power series expansions*.

$$(a + b)^r = a^r + \frac{r}{1!} a^{r-1}b + \frac{r(r-1)}{2!} a^{r-2}b^2 + \ldots \qquad e^x = 1 + \frac{x}{1!} + \frac{x^2}{2!} + \frac{x^3}{3!} + \ldots$$

$$(1 + x)^r = 1 + rx + \frac{r(r-1)}{2!} x^2 + \ldots$$

$$\ln(1 \pm x) = \pm x - \frac{1}{2}x^2 \pm x^3 + \ldots$$

(In the preceding expressions that involve r, r is a negative integer or a fraction.)

It is often desirable *to approximate* the solution of a problem, either because the exact solution may be difficult to obtain or may not be necessary or because we may need the value for a limiting case, say, when the variable x is very small. The preceding power series expansions are very helpful for such approximations. For example, when x is very small ($x \ll 1$), the x^2 and other higher-order terms are negligibly small and

$$(1 + x)^r \approx 1 + rx$$

$$e^x \approx 1 + x,$$

$$\ln (1 \pm x) \approx \pm x$$

Example 3.19.1

Use the power series approximation to find the square root of 1.008.

$$\sqrt{1.008} = (1.008)^{1/2}$$

$$= (1 + 0.008)^{1/2}$$

Use the approximation $(1 + x)^r \approx 1 + rx$. $\qquad \approx 1 + \frac{1}{2}(0.008)$

$$= 1 + 0.004 = 1.004$$

Example 3.19.2

Show that $(x \pm \delta x)^r = x^r \pm rx^{r-1} \, \delta x$, where δx is very small compared to x.

$$(x \pm \delta x)^r = x^r \, (1 \pm \delta x/x)^r$$

Use the approximation $(1 + x)^r \approx 1 + rx$.

$$= x^r \, (1 \pm r\delta x/x)$$

$$= x^r \pm x^r \, r \, \delta x/x$$

$$= x^r \pm rx^{r-1} \, \delta x$$

This relation is used to detrmine the total uncertainty (see Table 10.1) when a quantity is raised to a power.

Example 3.19.3 (from special relativity)

The ability do work is called energy. Kinetic energy is the energy of motion. For a spaceship moving at a speed v, its kinetic energy is given by $K = mc^2 \left(\dfrac{1}{\sqrt{1 - v^2/c^2}} - 1 \right)$, where c is the speed of light. Show that, for small velocity (v

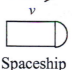

$\ll c$), kinetic energy of the spaceship becomes $\dfrac{1}{2} mv^2$.

Spaceship

We have

$$K = mc^2 \left(\frac{1}{\sqrt{1 - v^2/c^2}} - 1 \right) = mc^2 \left[\left(1 - v^2/c^2\right)^{-1/2} - 1 \right]$$

Now, $(1 + x)^r \approx 1 + rx$; therefore, $\left(1 - v^2/c^2\right)^{-1/2} \approx 1 - \left(-\dfrac{1}{2} \right) \dfrac{v^2}{c^2}$ $[x = -v^2/c^2$ and $r = -$

½].

Thus, the preceding expression for kinetic energy can be written as

$$K = mc^2 \left(\frac{1}{\sqrt{1 - v^2/c^2}} - 1 \right)$$

$$\approx mc^2 \left(\left(1 - \left(-\frac{1}{2} \right) \frac{v^2}{c^2} \right) - 1 \right)$$

$$= mc^2 \frac{1}{2} \frac{v^2}{c^2} = \frac{1}{2} mv^2$$

Exercises

46. Use the power series approximation to find the following:

(a) $(1.009)^{1/3}$ (b) $\ln(1.04)$ (c) $(0.98)^{1/2}$

47. According to the relativistic time dilation formula, if a car travels at speed v with respect to the earth, the time t_0 recorded by the driver according to his own watch is related to the time t recorded by an observer on the earth by

$t = \dfrac{t_0}{\sqrt{1 - v^2/c^2}}$, where $c = 3 \times 10^8$ m/s is the speed of light. If the speed of

the car is 110 km/h, and the driver travels for 10 seconds according to his watch, show that the difference between t and t_0 is about 5×10^{-14} seconds.
[*Hint*: First, show that the relation can be approximated as $t = t_0 (1 + \frac{1}{2} v^2/c^2)$. Also, you must convert the speed of the car to meters per second.]

48. According to the general theory of relativity, the speed of light is reduced in the space surrounded by an astronomical body (such as a star) of mass M and radius r by the factor $\sqrt{1 - 2GM/rc^2}/\sqrt{1 + 2GM/rc^2}$ in comparison with its speed, c, in free space. Here G is a constant, called the Gravitational constant. Show that the preceding factor can be approximated as $1 - 2GM/rc^2$ when $2GM/rc^2$ is much less than 1.

49. The value of acceleration due to gravity g' at a height h from the earth's surface is given by $g' = GM/(r + h)^2$, where M and r are the mass and the radius of the earth, respectively, and G is a constant (called universal gravitational constant). If g is the acceleration due to gravity on the surface of the earth ($h = 0$ on the surface), show that when $h \ll r$, $(g' - g) \approx -2gh/r$. [*Hint*: $(r + h)^2 = r^2 (1 + h/r)^2$.]

*3.20 Complex Numbers

Solutions of some quadratic equations are not real numbers. They involve the square root of negative numbers such as $\sqrt{-2}$. Since there is no real number whose square is –2, mathematician Rene Descartes (1597–1650) called them *imaginary numbers*. Leonard Euler (1707–1783) first used the letter i for $\sqrt{-1}$. Thus, $i^2 = -1$. Similarly, other powers of i can be determined and the results vary between $+i$, -1, $-i$, 1, as shown in Figure 3.10.

Imaginary numbers are important in physics. For example, the behavior of alternating current in electricity can be described conveniently using imaginary numbers. Complex numbers are useful in solving differential equations, such as in mechanics to study the behavior of a particle of mass that moves under the influence of an elastic force. In relativity theory, time is considered to be an imaginary fourth dimension that has been added to the three space dimensions and complex numbers are often used in the theory.

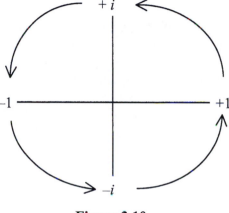

Figure 3.10

The imaginary numbers are a subset of what are called *complex numbers*. A complex number is a number that can be expressed in the form $a + bi$ ($= z$, say), where a and b are real numbers and $i = \sqrt{-1}$. The number a is called the real part and b is called the imaginary part of the complex number. The complex numbers $a + bi$ and $a - bi$ are called *complex conjugates* of each other. The absolute value of $a + bi = |a + bi| = |z| = r = \sqrt{a^2 + b^2}$.

Remember the following properties for complex numbers:

(i) The complex numbers $a + bi$ and $c + di$ are equal if and only if $a = c$ and $b = d$.

(ii) The product of two complex numbers: $(a + bi)(c + di) = (ac - bd) + (ad + bc)i$.

(iii) The division of two complex numbers: $\dfrac{(a+bi)}{(c+di)}$.

Multiply both numerator and denominator by $(c-di)$.

$$\frac{(a+bi)}{(c+di)} = \frac{(a+bi)(c-di)}{(c+di)(c-di)}$$

Rearrange the terms.

$$= \frac{(ac+bd)+(bc-ad)i}{c^2+d^2}$$

We often represent a complex number $z = a + bi$ by a point on a two-dimensional complex plane, so that the real part of z represents the x-coordinate and the imaginary part of z represents the y-coordinate of the point in the plane. The x-axis is known as the real axis and the y-axis is known as the imaginary axis (Figure 3.11).

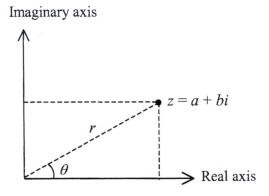

Figure 3.11

A complex number is also expressed in trigonometric form [or polar form with coordinates (r, θ)]. This is discussed in Section 5.8 of Chapter 5.

Example 3.20.1

Two complex numbers are given as, $z_1 = 3 + 4i$ and $z_2 = 4 - 2i$.

(a) What are the complex conjugates of z_1 and z_2?

(b) What are the absolute values of z_1 and z_2?

(c) Determine $z_1 + z_2$, $z_1 \cdot z_2$ and z_1/z_2.

(a) The complex conjugates are $3 - 4i$ and $4 + 2i$.

(b) The absolute values are $|z_1| = \sqrt{3^2 + 4^2} = 5,$ $|z_2| = \sqrt{4^2 + (-2)^2} = 2\sqrt{5}.$

(c) $z_1 + z_2 = (3 + 4i) + (4 - 2i) = (3 + 4) + (4 - 2)i = 7 + 2i$

$z_1 \cdot z_2 = (3 + 4i)(4 - 2i) = [3 \times 4 - 4 \times (-2)] + [4 \times 4 + 3 \times (-2)]i$

$$= 20 + 10i$$

$$\frac{z_1}{z_2} = \frac{3 + 4i}{4 - 2i}$$

Multiply both numerator and denominator by $(4 + 2i)$.

Thus,

$$\frac{z_1}{z_2} = \frac{(3 + 4i)(4 + 2i)}{(4 - 2i)(4 + 2i)} = \frac{(3 \times 4 - 4 \times 2) + (4 \times 4 + 3 \times 2)i}{4^2 + 2^2}$$

$$= \frac{4 + 22i}{20} = 0.2 + 1.1i$$

Exercises

50. Two complex numbers are given as $z_1 = -1 + \sqrt{3}\,i$ and $z_2 = 2 + 2\sqrt{3}\,i.$

    (a) What are the complex conjugates of z_1 and z_2?

    (b) What are the absolute values of z_1 and z_2?

    (c) Determine $z_1 + z_2$, $z_1 \cdot z_2$ and z_1/z_2.

51. Express the following complex numbers in the form $z = a + bi$:
    (a) $i^7 + i^2$ (b) $z = (2 + 3i)(2 - i)(4 + 5i)$

    (c) $\left(\dfrac{2 + 3i}{4 + 2i} \right)^2$ (d) $\dfrac{1 + 2i}{2 - 3i} + \dfrac{4 - 6i}{2 + 3i}$

Answers to Exercises in Chapter 3

2. 2.3×10^7 light-years $\qquad$ 3. 210 N $\qquad$ 4. 2.4×10^5 pascals

5. 1.8×10^4 N/m^2 $\qquad$ 6. 2.7×10^4 m/s $\qquad$ 7. 1.8 lux

8. 0.15 times the value at the surface $\qquad$ 9. 4.7 $\qquad$ 10. $x^{-4}y^8z^{-7}$

11. 1 $\qquad$ 12. $x^7y^{-11}z^3$ $\qquad$ 13. $64\,a^{-9}b^{-6}c^{-6}$

14. (a) $(t-3)(t-4)$ $\qquad$ (b) $2(x+4)(x-6)$ $\qquad$ (c) $6a(b+4)^2$

(d) $(2a+3)(a+1)$ $\qquad$ 15. $x = -2$ $\qquad$ 16. $x = 5$

17. $x = 5$ $\qquad$ 18. $G = Fr^2/Mm$ $\qquad$ 19. $P = RT/(V-b) - a/V^2$

20. $\alpha = (L - L_0)/(L_0\Delta T)$ $\qquad$ 21. $f = d_0d_i/(d_0 + d_i)$ $\qquad$ 22. 0.4 ohm

23. 26°C $\qquad$ 24. $x = 2$, $y = 1$ $\qquad$ 25. $x = 2$, $y = 3$

26. $x = 1$, $y = 1$ $\qquad$ 27. $x = 4$, $y = 2$ $\qquad$ 28. $x = 6$, $y = 4$

29. $x = 3$, $y = 4$ $\qquad$ 30. (a) 3.5, −3.5

(b) $t = \dfrac{1+\sqrt{10}}{3}, \dfrac{1-\sqrt{10}}{3}$ $\qquad$ (c) $y = \dfrac{-1+\sqrt{-3}}{2}, \dfrac{-1-\sqrt{-3}}{2}$ $\qquad$ 31. 2.9 s

32. (a) $x > -9$ $\qquad$ (b) $x > -7/3$ $\qquad$ (c) $-1/3 < x \le 13/3$ $\qquad$ 33. 95 or more

34. (a) $3\log_b x + 2\log_b y + 10\log_b z$ $\qquad$ (b) $\dfrac{2}{3}\log_b x + \log_b y - \dfrac{2}{3}\log_b z$

(c) $\log_b x + 2\log_b y$

35. (a) $\log_b\left[x^4 y^{1/2}\Big/\left(x+y\right)^2\right]$ $\qquad$ (b) $\log_b[x(x+y)^2]$ $\qquad$ 36. 10^{-7} W/m^2

37. 1 W/m^2 $\qquad$ 38. 34 dB $\qquad$ 39. 1.0×10^{-7} gram-ions per liter

40. 6.3 $\qquad$ 41. 4.8 on the Richter scale

42. $\lambda = 0.15$ day^{-1} $\qquad$ 43. 6.2 s $\qquad$ 44. 1 ms

45 (i) $= a^5 + 5a^4b + 10a^3b^2 + 10a^2b^3 + 5ab^4 + b^5$ (ii) $a^4/81 - 2a^3b/27 + a^2b^2/6 - ab^3/6 + b^4/16$

46 (a) 1.003 $\qquad$ (b) 0.04 $\qquad$ (c) 0.99. $\qquad$ 50 (a) $-1-\sqrt{3}\,i$, $2 - 2\sqrt{3}\,i$

(b) $\left|z_1\right| = 2$, $\left|z_2\right| = 4$ $\qquad$ (c) $z_1 + z_2 = 1 + 3\sqrt{3}\,i$, $z_1 \cdot z_2 = -8$, $z_1/z_2 = (1 + \sqrt{3}\,i)/4$

51. (a) $-1-i$ $\qquad$ (b) $8 + 51i$ $\qquad$ (c) $(33/100) + (28/50)i$ $\qquad$ (d) $(-1/13)(14 + 17i)$

Chapter 4
Geometry: Shapes and Plots

The word *geometry* means the measurement of earth. Many of the early discoveries in geometry were motivated by the need for distance and area measurements of the earth. In general, geometry deals with *the property and measurement of regular shapes* including lines, angles, curves, planes, and surfaces. The rules of geometry were first studied systematically by the Greek mathematician Euclid (300 B.C.). In physics, you will often deal with regular shapes and curves. For example, the shapes of the earth and the sun can be approximated as spheres; the orbit for the motion of an electron in a normal hydrogen atom can be approximated as a circle; in absence of air resistance, the path for the motion of any projectile (such as a soccer ball) is a parabola and so on. Knowledge of geometry is extremely useful in studying such shapes and the physical situations involving them.

Angles, Triangles, and Quadrilaterals
4.1   Angles

In plane geometry, also called Euclidean geometry, *angles* are formed when two straight lines intersect on a plane (Figure 4.1). The point of intersection is called the called the *vertex* of the angle. Figure 4.2 shows an angle θ formed by an arc s of a circle of radius r at its center. Here the two radii intersect at the center, the vertex of the angle.

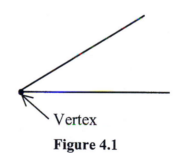

Vertex

Figure 4.1

An angle is considered to be positive if it is measured counterclockwise from the positive x-axis, and it is negative if measured clockwise from the positive x-axis.

An angle is measured in *degrees* or in *radians*. The most common unit of measure for angle is the *degree*. If the circumference of a circle is divided equally into 360 parts, each part would make a 1-degree angle at the center of the circle. The ratio of the length of the circular arc, s, to the radius r of a circle is the measure of the angle in *radians*

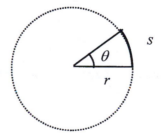

Figure 4.2

subtended by the arc at the center of the circle. From Figure 4.2,

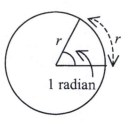

$$\theta = \frac{s}{r} \text{ in radians.}$$

Figure 4.3

From the preceding equation, when $s = r$, θ is 1 radian. Thus, a *radian* is the angle subtended at the center of a circle by an arc whose length is equal to the radius of the circle (Figure 4.3). Since the circumference of a circle is $2\pi r$, a complete circle has an angular measure of 2π radians, also called 360 degrees (°). Thus, $360° = 2\pi$ radians or $1° = \pi/180$ radian. Hence, 1 radian $= 180/\pi = 57.3°$, and, further,

$45° = \pi/4$ radians	$90° = \pi/2$ radians
$180° = \pi$ radians	$270° = 3\pi/2$ radians
$360° = 2\pi$ radians	

In astrophysics, when talking about stars and galaxies, we often measure angles much less than a degree and use *seconds of arc* to measure angles. A degree is divided into 60 minutes of arc, each of which is further divided into 60 seconds of arc. Thus, $1° = 3600$ seconds of arc. 1 radian $= 57.3°$ degrees $= 57.3 \times 3600$ arcseconds $= 2.06 \times 10^5$ arcseconds.

Example 4.1.1

The distance between the sun and the earth is 1.5×10^{11} m. The sun makes an angle of about 0.4° to us on the earth. Use this information to estimate the radius of the sun.

List the data. Change the angle to radians.

$$\text{Distance, } r = 1.5 \times 10^{11} \text{ meters}$$

$$\theta = 0.4° = 0.4 \times \pi/180 \text{ radian}$$

For small angles, the arc length and the chord length (the diameter of the sun, in this example) are approximately equal. Use the relation $s = r \cdot \theta$ to determine the diameter of the sun.

$$s = r \cdot \theta$$

Substitute the values of r and θ. $s = (1.5 \times 10^{11}$ m$) \times 0.4 \times \pi/180$ radian$)$

$$= 10.5 \times 10^8 \text{ m} = 1.1 \times 10^9 \text{ m}$$

Exercise

1. A bird can distinguish objects that subtend an angle of at least 0.02° to its eye. What is the angle in radians? When the bird is flying at a height of 120 m, how small an object can it distinguish?

Bird's eye

0.02°   120 m

?

Classification of Angles

When two lines meet to form one straight line, it makes an angle of 180° (Figure 4.4a). If an angle is 90°, it is called a *right angle* (Figure 4.4b).

(a)   180°

This mark indicates a right angle. **(b)**

Right angle = 90°

Figure 4.4

An angle is called *acute* if its value is between 0° and 90° (Figure 4.5a). An angle is called *obtuse* if its value is between 90° and 180° (Figure 4.5b).

(a) Acute angle

(b) Obtuse angle

Figure 4.5

If two angles sum to 90° (such as angles A and B in Figure 4.6a), they are called *complementary angles*. If two angles sum to 180° (such as angles A and B in Figure 4.6b), they are called *supplementary angles*.

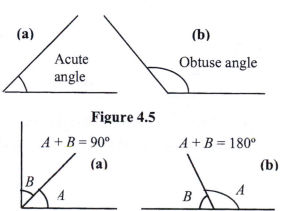

$A + B = 90°$ (a)   B   A

$A + B = 180°$ (b)   B   A

Figure 4.6

Equality of Angles

Some common situations in physics, when two angles can be equal, are shown in Figures 4.7a–d.

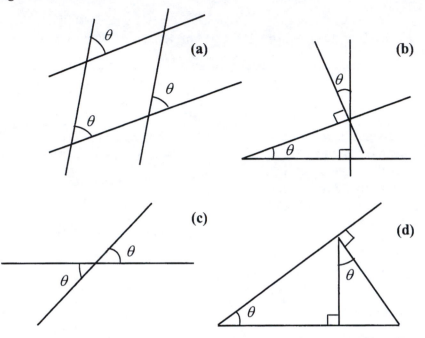

Figures 4.7

From Figures 4.7a–d, note that two angles are equal when any of the following are true:

(i)   The angles are made by a line crossing parallel lines, as shown in Figure 4.7a.

(ii)   The two sides of the angles are perpendicular, as shown in Figure 4.7b.

(iii) The angles are opposite angles at the intersection of two lines, as shown in Figure 4.7c.

(iv)   The angles have the same complement, as shown in Figure 4.7d.

4.2   Triangles and Quadrilaterals

When an object moves in three-dimensional space, it can move left or right, front or back, and up or down. However, in many circumstances, the motion of an object in space is confined to a two-dimensional plane. For example, when you throw a ball, its motion is confined to a plane; the motion of a planet around the sun and the motion of a satellite orbiting the earth are also confined to a plane. Two-dimensional geometry is useful to describe the physics of these objects. The two-dimensional figures that are common in physics, such as triangles and quadrilaterals, are described next.

Triangles

A *triangle* is a figure formed by the intersections of three straight lines on a plane. A triangle has three sides and three internal angles. The three sides and the three angles can be different. Also, it is possible that the two sides and two angles or all three sides and three angles be equal.

Equilateral Triangle: An equilateral triangle is a triangle that has equal sides and each angle equal to 60°, as shown in the Figure 4.8a.

Isosceles Triangle: An isosceles triangle is a triangle that has two equal sides and two equal angles, as shown in Figure 4.8b.

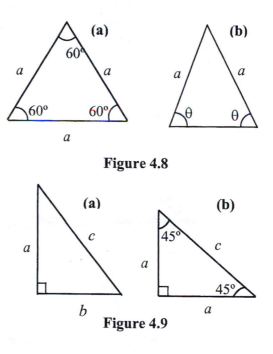

Figure 4.8

Right Triangle: A right triangle (Figure 4.9a) is a triangle that has one angle 90°. In a *right isosceles triangle* (Figure 4.9b), the acute angles are each 45°.

Figure 4.9

The length of the base, b, height, a, and hypotenuse, c, of a right triangle are related by the *Pythagorean theorem*:

$$c^2 = a^2 + b^2.$$

30–60–90 *Triangle*: If you split an equilateral triangle exactly in half, each part becomes a 30–60–90 triangle (Figure 4.10). That is, the angles of the resulting triangles are 30°, 60°, and 90°. In this case, the length of the side opposite the 30° angle is one-half the length of the hypotenuse.

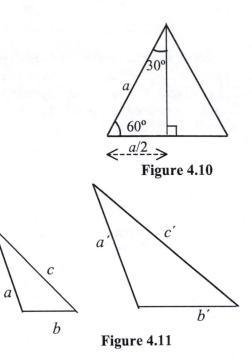

Figure 4.10

Similar Triangles: Similar triangles are triangles that have the same shapes but that differ only in size or orientation, as shown in Figure 4.11. The internal angles of two similar triangles are equal and the ratios of the corresponding sides are equal. If the sides are a, b, and c and a', b', and c', respectively, for the two similar triangles, then

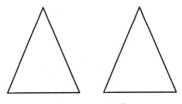

Figure 4.11

$$\frac{a}{a'} = \frac{b}{b'} = \frac{c}{c'}.$$

Congruent Triangles: Two triangles are said to be *congruent* if they have exact fit when one is placed on the top of the other. That is, congruent triangles have the same size and shape. The two triangles in Figure 4.12 are congruent.

Figure 4.12

Note that in any triangle (such as that shown in Figure 4.13),

(i)   The sum of the three interior angles is 180°. That is,

$$A + B + C = 180°.$$

Figure 4.13

(ii)   The exterior angle, θ, is the sum of the two opposite interior angles. That is, $A + B = \theta$.

Example 4.2.1

In the figure on the right, the points D and E are the middle points of AB and CB, respectively. The angles are shown in the figure. What are the values of the angles α, β, γ, and δ? If DE is 10 cm, what is the length AC?

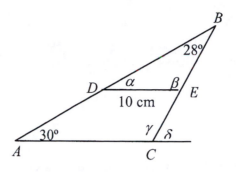

DE is parallel to AC.

$\alpha = 30°$ [since the two sides of the angles α and $30°$ are parallel]

$\beta = 180° - (30° + 28°)$ [summation of all angles of the triangle DBE is $180°$] $=122°$.

$\gamma = 180° - (30° + 28°)$ [summation of all angles of the triangle ABC is $180°$] $= 122°$.

Note that $\beta = \gamma$, as expected, since they have two sides parallel.

$\delta = 30° + 28° = 58°$ [exterior angle is the sum of two opposite interior angle of a

triangle].

The triangles ABC and DBE are similar. Since the ratio of the sides DB and AB is 1 : 2

(D is the middle point of AB), the ratio of the sides DE and AC is 1 : 2.

That is, AC $= 2DE = 2 \times 10$ cm $= 20$ cm.

Example 4.2.2

In the diagram to the right find the angles A, B, C, and D.

$B = 50°$, because the lines forming B are mutually perpendicular to the lines forming the angle $50°$.
$A = 90° - B = 90° - 50° = 40°$.
$D = B$ (vertical angles) $= 50°$.
The small triangle that contains D is a right-angled triangle.
$C = 90° - D = 90° - 50° = 40°$
(since C's vertical angle is complementary to D).

Example 4.2.3

Two tangent lines (AT_1 and AT_2) are drawn from a point A to a circle whose center is at O. The line OA makes an angle of 30° at A with the tangent line AT_1. Determine the angles T_2AO, AOT_1, AOT_2, and T_1OT_2 from the figure on the right.

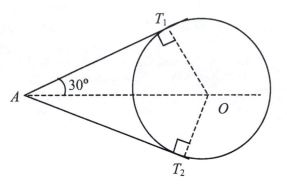

Note that a tangent line drawn to a circle at any point makes a right angle with the radius drawn from the point. That is, angles AT_1O and AT_2O are 90° each. From symmetry, it is clearly seen that the triangles T_1AO and T_2AO are identical. Thus, the angle T_2AO is also 30°.

Angle AOT_1 = 180° – (30° + 90°) [the sum of the three internal angles of the triangle AOT_1 is 180°]

$$= 180° – 120° = 60°.$$

Angle AOT_2 = 180° – (30° + 90°) (the sum of the three internal angles of the triangle AOT_2 is 180°)

$$= 180° – 120° = 60°.$$

Angle T_1OT_2 = Angle AOT_1 + Angle AOT_2 = 60° + 60° = 120°.

Exercise

2. For the triangle on the left, find the unknown angles A and B. A perpendicular has been drawn to one of the sides of the same triangle, as shown on the right. Determine the angles C and D.

3. Find the unknown angles and unknown sides of the triangle RST. The side PQ is parallel to side ST.

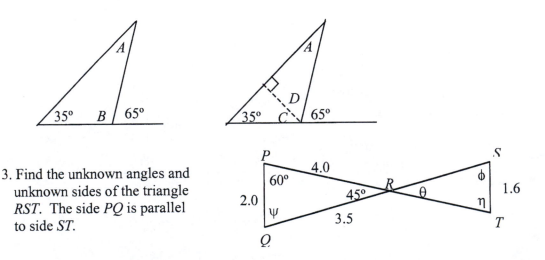

Quadrilaterals

When four straight lines intersect in a plane, they form a closed figure with four sides. This figure is called a *quadrilateral*. The following quadrilaterals, where the two opposite sides are parallel to each other, as shown in Figure 4.14, are common in physics. These are called *parallelogram*, *rectangle*, and *square*, respectively. The line (dotted line for in the case of the square) that joins the two opposite corners of a quadrilateral is called its *diagonal*. Each of the figures has two diagonals. The two diagonals of a parallelogram have different lengths. The two diagonals have the same length for a rectangle or a square because of the symmetry of these two figures.

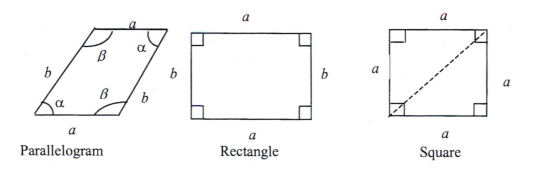

Parallelogram   Rectangle   Square

Figure 4.14

Note that
- A *parallelogram* has opposite pairs of equal sides and a pair of equal angles.
- A *rectangle* is a parallelogram whose angles are all 90°.
- A *square* is a rectangle whose four sides are equal.

The sum of the four angles of any quadrilateral is 360°. Joining a diagonal, as shown in Figure 4.15, can easily verify the preceding statement. The diagonal divides the quadrilateral into two triangles, for each of which the sum of the three angles is 180°.

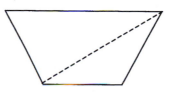

Figure 4.15

Shapes of Objects

4.3   Perimeter, Area, and Volume of Standard Shapes

A one-dimensional object (such as a very thin string) has a length, but no area or volume. A two-dimensional object (such as a piece of paper) has a perimeter and a surface area. A three-dimensional object (such as a solid metal cylinder) has area for each of its surfaces and it occupies a certain amount of space, called its volume. Perimeters, areas, and volumes of common figures are often needed for solving real-world problems.

Note: A perimeter has a dimension $[L]$; that is, it has the first power in length. An area has a dimension $[L^2]$; that is, it has the second power of length; a volume has a dimension $[L^3]$, the third power of length.

It is worthwhile to remember the actual formulas for the perimeter, area, and volume of the common figures shown in Figure 4.16. These common figures include regular-shaped two-dimensional objects such as a circle and an ellipse and three-dimensional solid objects such as a rectangular parallelepiped, a cube, a sphere, and a right circular cylinder.

Two-Dimensional Figures

Figure 4.16

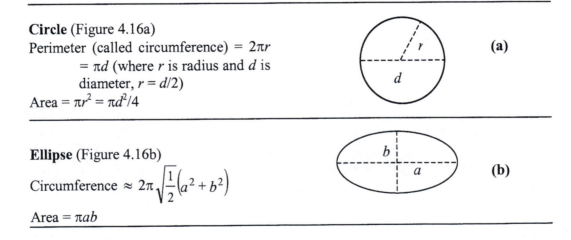

Circle (Figure 4.16a)
Perimeter (called circumference) = $2\pi r$
      = πd (where r is radius and d is
      diameter, $r = d/2$)
Area = $\pi r^2 = \pi d^2/4$

(a)

Ellipse (Figure 4.16b)

Circumference $\approx 2\pi\sqrt{\dfrac{1}{2}\left(a^2 + b^2\right)}$

Area = πab

(b)

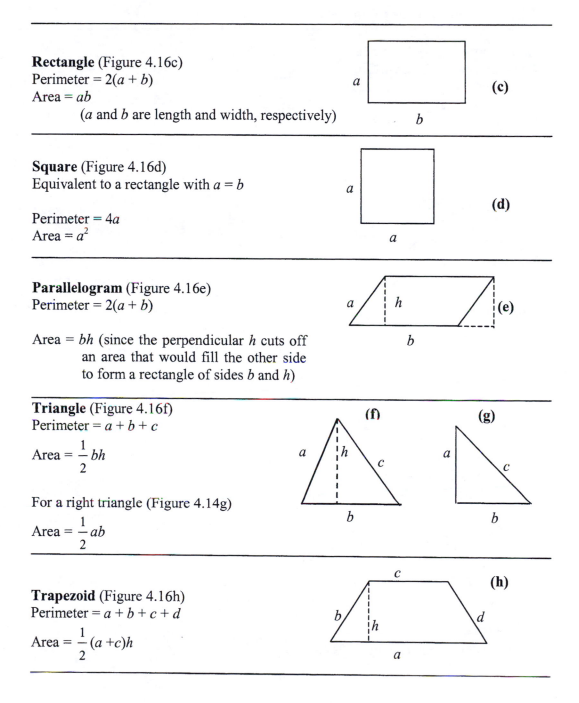

Rectangle (Figure 4.16c)
Perimeter $= 2(a + b)$
Area $= ab$
  (a and b are length and width, respectively)

(c)

Square (Figure 4.16d)
Equivalent to a rectangle with $a = b$

Perimeter $= 4a$
Area $= a^2$

(d)

Parallelogram (Figure 4.16e)
Perimeter $= 2(a + b)$

Area $= bh$ (since the perpendicular h cuts off
   an area that would fill the other side
   to form a rectangle of sides b and h)

(e)

Triangle (Figure 4.16f)
Perimeter $= a + b + c$

Area $= \dfrac{1}{2}bh$

For a right triangle (Figure 4.14g)

Area $= \dfrac{1}{2}ab$

(f)   **(g)**

Trapezoid (Figure 4.16h)
Perimeter $= a + b + c + d$

Area $= \dfrac{1}{2}(a + c)h$

(h)

Three-Dimensional Figures

Rectangular parallelepiped (Figure 4.17a) and **cube** (Figure 4.17b)
Surface area = sum of the surface areas of all six surfaces
For a rectangular parallelepiped,
Surface area $= 2(ab + bc + ca)$

Volume $= abc$

For a cube, $a = b = c$

Surface area $= 6a^2$
Volume $= a^3$

(a)   **(b)**

Sphere (Figure 4.17c)
Surface area $= 4\pi r^2$

Volume $= \dfrac{4}{3}\pi r^3 = \dfrac{1}{6}\pi d^3$, where $d = 2r$

(c)

Right circular cylinder (Figure 4.17d)
Surface area = area of the two circular end caps plus the area of the cylindrical wrapper $= 2\pi r^2 + 2\pi rL$

Volume = the volume generated by sweeping the cross-sectional area a distance L along the axis $= \pi r^2 L$

(d)

Right circular cone (Figure 4.17e)

Volume $= \dfrac{1}{3}\pi r^2 h$

(e)

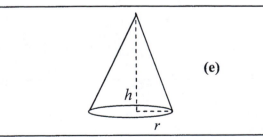

Example 4.3.1
(a)   Calculate the circumference and area of a circle of radius 0.10 m.
(b)   Calculate the area of a triangle of height 0.20 m and base 1.40 m.
(c)   Calculate the surface area and volume of a right circular cylinder of radius 0.20 m and length 0.40 m.
(d)   Determine the volume of a cone of radius 0.20 m and height 0.80 m.

This problem is straightforward. Use the proper formula for the circumference, area, and volume to solve the problem.

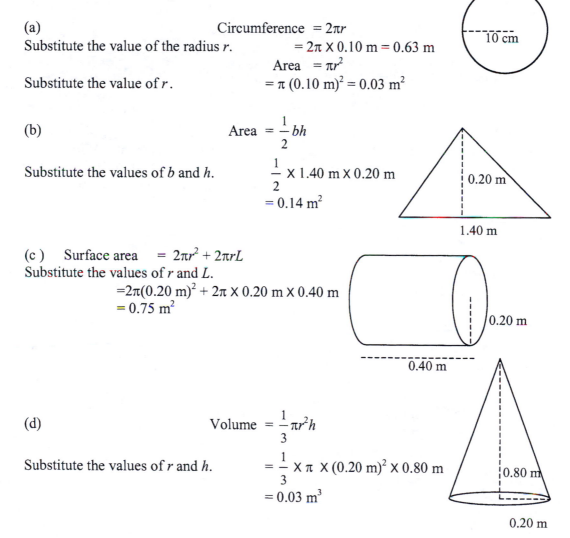

(a)   Circumference $= 2\pi r$
Substitute the value of the radius r. $= 2\pi \times 0.10 \text{ m} = 0.63 \text{ m}$
                                      Area $= \pi r^2$
Substitute the value of r. $= \pi (0.10 \text{ m})^2 = 0.03 \text{ m}^2$

(b)   Area $= \dfrac{1}{2} bh$

Substitute the values of b and h. $\dfrac{1}{2} \times 1.40 \text{ m} \times 0.20 \text{ m}$

                                      $= 0.14 \text{ m}^2$

(c)   Surface area $= 2\pi r^2 + 2\pi rL$
Substitute the values of r and L.
                      $= 2\pi (0.20 \text{ m})^2 + 2\pi \times 0.20 \text{ m} \times 0.40 \text{ m}$
                      $= 0.75 \text{ m}^2$

(d)   Volume $= \dfrac{1}{3} \pi r^2 h$

Substitute the values of r and h. $= \dfrac{1}{3} \times \pi \times (0.20 \text{ m})^2 \times 0.80 \text{ m}$

                                      $= 0.03 \text{ m}^3$

Note that, in the answers, circumference has unit m, surface area has unit m², and volume has unit m³, as expected.

Example 4.3.2

The earth can be approximated to be a sphere of average radius 6.38×10^6 m. Determine the volume of the earth.

Radius $r = 6.38 \times 10^6$ m.

Use the formula for the volume of a sphere, which is $\frac{4}{3}\pi r^3$.

$$\text{Volume of the earth} = \frac{4}{3}\pi r^3$$

Substitute the value of the radius, r. $\quad = \frac{4}{3}\pi (6.38 \times 10^6 \text{ m})^3 = 1.09 \times 10^{21} \text{ m}^3$

Exercises

4. Density. The average density D of a substance is defined as the ratio of its mass M to its volume V. Thus, $D = M/V$. If the mass of the earth is 5.97×10^{24} kg and its average radius is 6.38×10^6 m, determine the average density of the earth.

5. A circle is inscribed in a square for which sides are each 2.0 cm long and are tangents to the circle, as shown on the right. Determine the area of each corner segment.

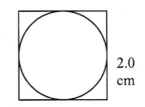

2.0 cm

6. A spherical shell of copper, as shown on the right, has an inner diameter of 2 cm and an outer diameter of 2.8 cm. Determine the volume of copper in the shell. (*Hint*: The volume of the spherical shell is equal to the difference between the volumes of the two spheres.)

2.8 cm

2.0 cm

The Coordinate Systems

4.4   The Cartesian Coordinate System

In geometry, coordinate systems are used to

- Find the location of an object with respect to a suitably chosen origin,

- Find the distance between two objects,

- Determine the trajectory of motion,

- And more.

Two Dimensions

In a Cartesian coordinate system (Figure 4.18) the two coordinate axes x and y are perpendicular to each other and the divisions of the x- and y-axes are equally spaced. The x- and y-axes meet at a point called the *origin*. A point P is labeled by (x, y), where the magnitude of x is the distance of the point measured from the y-axis and the magnitude of y is the distance of the point measured from the x-axis. The values x and y can be positive or negative, depending on whether the distances are toward positive or negative axis from the origin.

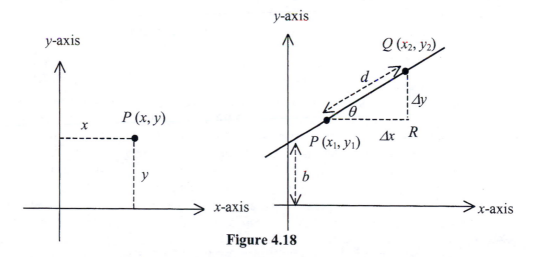

Figure 4.18

Distance between Two Points and the Equation of a Straight Line

Let (x_1, y_1) and (x_2, y_2) be the Cartesian coordinates of two points P and Q, respectively. The distance d between the points P and Q is the hypotenuse of the right triangle PQR in Figure 4.18 and can be obtained by using the Pythagorean theorem. Thus,

$$d = \sqrt{\Delta x^2 + \Delta y^2} = \sqrt{(x_2 - x_1)^2 + (y_2 - y_1)^2}.$$

Here,

$$\Delta x = x_2 - x_1 \quad \text{and} \quad \Delta y = y_2 - y_1.$$

The line joining P and Q is called the *Cartesian graph*, and the equation of the straight line joining the two points P and Q is given by

$$y = mx + b,$$

where $m = \Delta y / \Delta x = \tan \theta$ (θ is the angle made by the line with the positive x-axis) and b is the length of the intercept of the line with the y-axis at the origin ($x = 0$). The steepness of the line is measured by the ratio m and is called the *slope* of the straight line. If the line passes through the origin, then $b = 0$, and the equation of the line is $y = mx$.

When a relationship between two variables can be written as $y = mx + b$, the relationship is called *linear*. A linear relationship between physical variables is common in physics. For example, when heated, a solid increases its length, and the relationship between the length L and temperature rise ΔT is approximately linear. The relationship is represented by $L = L_0 + L_0 \alpha \Delta T$, where L_0 is the constant initial length of the solid and α is a constant for the solid, called the *coefficient of linear expansion*. Comparing this relation with $y = mx + b$ reveals that here $x = \Delta T$, $y = L$, $m = L_0 \alpha$, and $b = L_0$.

Three Dimensions

An object can move in any direction in space. That is, it can move left or right, front or back, and up or down. To locate a point in space, three coordinates are needed. In a rectangular Cartesian system, the three axes labeled as x, y, and z are mutually perpendicular, and they intersect at one point: the origin O (Figure 4.19). The position of a point P is labeled by three numbers (x, y, z), where x is the perpendicular of the point from the y-z plane, y is the distance from the x-z plane, and z is the distance from x-y plane. If (x_1, y_1, z_1) and (x_2, y_2, z_2) are the Cartesian coordinates of two points, say, P and Q, respectively, then the distance between the points is given by

$$d = \sqrt{\Delta x^2 + \Delta y^2 + \Delta z^2} = \sqrt{(x_2 - x_1)^2 + (y_2 - y_1)^2 + (z_2 - z_1)^2}.$$

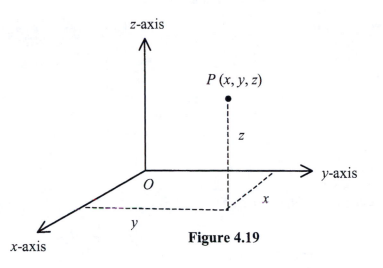

Figure 4.19

Here,

$$\Delta x = x_2 - x_1, \qquad\qquad \Delta y = y_2 - y_1, \quad \text{and} \quad \Delta z = z_2 - z_1.$$

The idea of the Cartesian coordinate system has been extended to include more than three dimensions. For example, in the special theory of relativity, time becomes the fourth dimension and the time axis is considered mutually perpendicular to the x-, y-, and z-axes. Four-dimensional geometry is used in the theory of relativity.

There are two types of three-dimensional angular coordinate systems that are used in physics. These are called the *spherical polar coordinate* system and the *cylindrical coordinate* system, which are described in Section 4.7.

Example 4.4.1

Determine the slope of a line that passes through the points (–2.0, 0.0) and (4.0, –4.0). Also, determine the angle θ that the line makes with the positive x-axis.

Use the expression for the slope to solve this problem.

$$\text{Slope, } m \;=\; \frac{\Delta y}{\Delta x} = \frac{y_2 - y_1}{x_2 - x_1}.$$

Substitute the values.

$x_1 = -2.0, \, y_1 = 0.0, \, x_2 = 4.0, \, y_2 = -4.0$
$$= \frac{-4.0 - 0.0}{4.0 - (-2.0)} = \frac{-4}{6} = -0.67$$

or $\tan \theta \;\; = -0.67$ (since $m = \tan \theta$)

Take inverse tan to get θ. $\theta = \tan^{-1}(-0.67) = -34°$

Example 4.4.2

Determine the equation of a straight line whose slope is 2.0 and that intersects the y-axis at a distance of -1.0 unit. Make a Cartesian plot of the line.

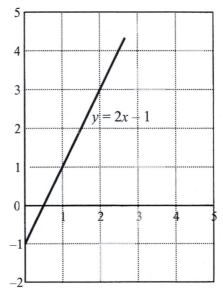

In Cartesian coordinates, the equation

of a line is $y = mx + b$.

In this example, $m = 2.0$ and $b = -1.0$.

Thus, the equation of the line is $y = mx$

$+ b = 2x - 1$.

The plot of the equation is shown on the right.

Example 4.4.3

Determine the magnitude of the displacement of a person whose initial and final locations in the Cartesian system are (2.0 m, 4.0 m) and (4.0 m, 6.0 m).

The magnitude of the displacement is the straight-line distance between the final and initial locations.

Thus, the magnitude of the displacement $= \sqrt{(x_2 - x_1)^2 + (y_2 - y_1)^2}$.

Substitute the values $= \sqrt{(4.0\,\text{m} - 2.0\,\text{m})^2 + (6.0\,\text{m} - 4.0\,\text{m})^2}$

$= 2.8\,\text{m}^2$

4.5   The Polar Coordinate System

The Cartesian coordinate system is most commonly used in introductory physics. However, it is not the most convenient system in all situations. In some situations, the polar coordinate system is more convenient. In the polar coordinate system, the coordinates of a point P are (r, θ), as shown in Figure 4.20. In this case, r is the radial distance of the point P from the origin O, and θ is the angle measured counterclockwise from the x-axis to r.

If (x, y) are the Cartesian coordinates of the point P [Figure 4.20], then*

$x = r \cos \theta$

$y = r \sin \theta.$

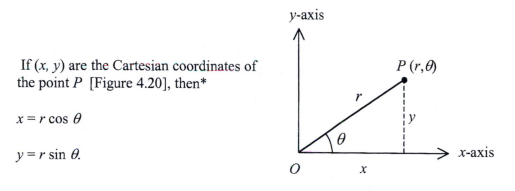

Figure 4.20

The polar coordinates r and θ can also be expressed in terms of x and y. Squaring and adding the preceding equations for x and y yields

$$x^2 + y^2 = r^2\cos^2\theta + r^2\sin^2\theta = r^2\,(\cos^2\theta + \sin^2\theta) = r^2$$

or $r = \sqrt{x^2 + y^2}.$

Also, dividing the expressions for x and y, we get

$$\frac{y}{x} = \frac{r\sin\theta}{r\cos\theta} = \frac{\sin\theta}{\cos\theta} = \tan\theta$$

or $\theta = \tan^{-1}\left(\dfrac{y}{x}\right).$

* Chapter 5 defines sin, cos, and tan trigonometric functions and gives details about these trigonometric functions and their properties.

Example 4.5.1
The Cartesian coordinates of a point are (4.0, 2.0). Determine the coordinates in the polar system.

In this example, $x = 4$ units and $y = 2$ units. The polar coordinates are (r, θ). Use the preceding two equations to determine r and θ. We have

$$r = \sqrt{x^2 + y^2} = \sqrt{4^2 + 2^2} = \sqrt{20} = 4.5 \text{ units}$$

$$\theta = \tan^{-1}\left(\frac{y}{x}\right) = \tan^{-1}\left(\frac{2}{4}\right) = 0.5 \text{ radian or } 27°.$$

4.6   Choosing a Coordinate System

The choice of a coordinate system is important to solve problems in physics. The choice may determine whether a problem can be solved easily or whether it will be difficult (or impossible) to solve.

In most situations in introductory physics, the Cartesian coordinate system is most convenient. In some situations, such as when there is *radial symmetry*, the polar coordinate system may be preferred.

Note that the origin of the coordinate system can be assumed to be at any point at your convenience. Think about what should be the logical location for the origin. In many situations, the initial location of the object of interest will be the logical choice of the origin. For motion of objects in the vertical direction, a good choice of the origin will be on the ground in most situations. For two-dimensional problems, such as the problems involving the motion of a projectile, use the horizontal direction for the x-axis and the vertical direction for the y-axis, and adopt a sign convention for x and y.

Most physics texts follow the sign convention that the *motion to the right* along a horizontal line is *positive* and motion in the *upward direction* is *positive*, as shown in Figure 4.21. Once a sign convention is selected, stick with it for the entire problem.

Note that the choice of the *x*-axis to be the horizontal direction and the *y*-axis to be the vertical direction may *not always* be the right choice.

For the motion of an object on an inclined surface, it is more convenient to choose the *x*-axis to be downhill along the inclined surface (with the downhill direction as the positive direction) and the *y*-axis to be perpendicular to the inclined surface (with the positive direction away from the surface), as shown in Figure 4.22. For this choice of axes, the motion of the object along the surface is restricted only to the *x*-direction, thus making the solution simpler.

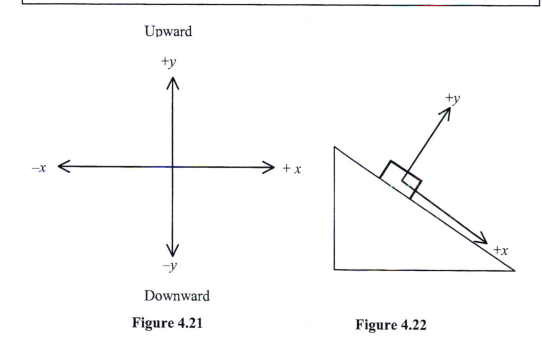

Upward

Downward

Figure 4.21   **Figure 4.22**

Note that the result of a problem or the physics of a problem does not depend on the choice of the coordinate system or the sign convention. Your result should be the same regardless of the choice of axes and the choice of sign convention.

Example 4.6.1

A skier is accelerating down a 30° hill at 2.0 m/s². She starts from rest and takes 20.0 seconds to reach the bottom of the hill. Determine the length of the incline.

In this case, the most convenient coordinate system is the Cartesian coordinate system, with the x-axis along the inclined surface and downhill as the positive direction. Choose the origin to be the initial location of the skier. Thus, as the skier accelerates downhill, she moves along the positive x-axis starting from the origin. (See the diagram on the right.)

Here, x_0 (the initial location of the skier) = 0.0 m.

v_0 = initial velocity = 0.0 m/s,

$a = 2.0$ m/s² ,

$t = 20.0$ s,

$x = ?$

To solve this problem, we apply the kinematic equation of physics that applies to the x-direction when the acceleration is constant. The equation is

$$(x - x_0) = v_0\, t + \frac{1}{2} at^2.$$

Substitute these values in the preceding equation to solve for x, the length of the incline.

$$x - 0 \quad = 0 \cdot 20.0 + \frac{1}{2}(2.0 \text{ m/s}^2)(20.0 \text{ s})^2,$$

$$x = 400 \text{ m}$$

4.7   The Spherical Polar and the Cylindrical Coordinate Systems

There are two types of three-dimensional angular coordinate systems that are commonly used:

- the *spherical polar coordinate* system, and

- the *cylindrical coordinate* system

The *spherical polar system* is very useful for situations that have spherical symmetry, such as in specifying the point on a sphere. In physics, you will encounter situations having spherical symmetry, such as situations involving a spherical mass or a spherical charge distribution. In the spherical polar coordinate system, a point is specified by r, θ, and ϕ and the point is written as (r, θ, ϕ). Here, r is the radial distance of the point from the origin to the point P on the sphere, θ is the angle measured from the z-axis to the radius r, and ϕ is the angle in polar coordinates, as shown in Figure 4.23.

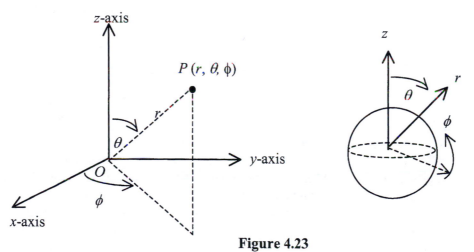

Figure 4.23

Note that r can vary from 0 to ∞; θ can vary from 0 to π (that is, positive z-axis to negative z-axis); ϕ can vary from 0 to 2π (moving on a circle on the xy-plane).

If (x, y, z) are the Cartesian coordinates of the point P, it can be shown that the relationships between the coordinates of Cartesian system and spherical polar system are given by

$$x = r \sin\theta \cos\phi$$
$$y = r \sin\theta \sin\phi$$

$$z = r\cos\theta$$

$$r = \sqrt{x^2 + y^2 + z^2} \qquad \theta = \cos^{-1}\left(\frac{z}{\sqrt{x^2 + y^2 + z^2}}\right) \qquad \phi = \tan^{-1}\left(\frac{y}{x}\right).$$

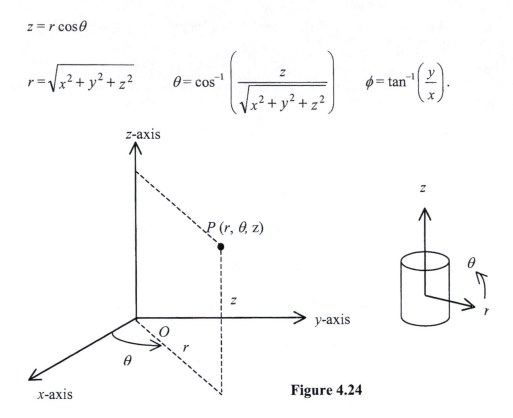

Figure 4.24

In the *cylindrical coordinate system* (Figure 4.24) a point is labeled by r, θ, and z and written as (r, θ, z), where r and θ are two polar coordinates (that is, these are defined in the same way as in the polar system) and the coordinate z is the z Cartesian coordinate (that is, defined in the same way as was defined for the z-coordinate in the Cartesian system). The cylindrical coordinate system is useful for three-dimensional systems where there is symmetry about one axis, say the z-axis (such as for a cylindrical object or a cylindrical shell).

The relationships between the coordinates of Cartesian system and cylindrical system are given by

$$x = r\cos\theta \qquad\qquad\qquad r = \sqrt{x^2 + y^2}$$

$$y = r\sin\theta \qquad\qquad\qquad \theta = \tan^{-1}\left(\frac{y}{x}\right)$$

$$z = z \qquad\qquad\qquad\qquad z = z.$$

Example 4.7.1

The position of an object in space in Cartesian system is given by the coordinates (3.0 m, 2.0 m, 1.0 m). Determine the coordinates of the position in the spherical polar system and in the cylindrical system.

In this example, $x = 3.0$ m, $y = 2.0$ m, and $z = 1.0$ m.

The coordinates of the point in the spherical polar system are

$$r = \sqrt{x^2 + y^2 + z^2}$$

Substitute the values.

$$= \sqrt{3^2 + 2^2 + 1^2} = \sqrt{14} = 3.7 \text{ m}$$

$$\theta = \cos^{-1}\left(\frac{z}{\sqrt{x^2 + y^2 + z^2}}\right)$$

Substitute the values.

$$= \cos^{-1}\left(\frac{1}{3.7}\right) = 1.4 \text{ radians}$$

$$\phi = \tan^{-1}\left(\frac{y}{x}\right)$$

Substitute the values.

$$= \tan^{-1}\left(\frac{2}{3}\right) = 0.6 \text{ radians.}$$

The coordinates of the point in the cylindrical system are

$$r = \sqrt{x^2 + y^2}$$

Substitute the values.

$$= \sqrt{3^2 + 2^2} = \sqrt{13} = 3.6 \text{ m}$$

$$\theta = \tan^{-1}\left(\frac{y}{x}\right)$$

Substitute the values.

$$= \tan^{-1}\left(\frac{2}{3}\right) = 0.6 \text{ radians}$$

$$z = 1.0 \text{ m}$$

Conic Sections

4.8   Conic Sections

The second-degree equations representing *circles*, *ellipses*, *parabolas*, and *hyperbolas* are common in physics. For example, the motion of a satellite around the earth is a circle, the path of motion of a planet around the sun is an ellipse, the path of a projectile (such as the motion of a soccer ball in air) in the absence of air resistance is a parabola, and the curve representing the pressure and volume of an ideal gas when its temperature is constant is a hyperbola. These figures were fully investigated in the seventeenth century by René Descartes (1596–1650) and Blaise Pascal (1623–1662). The circles, ellipses, parabolas, and hyperbolas are collectively called *conic sections*, because they can be created from the intersection of a plane with a cone, as shown in Figure 4.25. These curves are also defined geometrically and are distinguished by the algebraic equations that represent them.

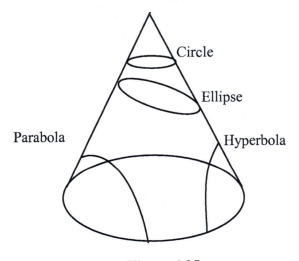

Figure 4.25

Circle

A *circle* is the simplest conic section. It is defined as the set of all points in the plane equidistant from a fixed point, called the *center* of the circle. The constant distance a is called the *radius* of the circle. If (x_0, y_0) are the coordinates of the center of a circle (Figure 4.26), then the equation of a circle is

$$(x - x_0)^2 + (y - y_0)^2 = a^2.$$

If the center of the circle is at the origin, the equation of a circle reduces to

$$x^2 + y^2 = a^2.$$

A circle is symmetric about all axes. A *tangent* line drawn at any point to a circle is always *perpendicular* to the *radius* of the circle. Circles have many practical applications, such as circular wheels and gears, merry-go-rounds, ferris wheels, and pizza cutters.

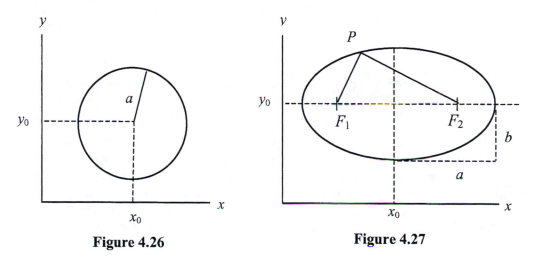

Figure 4.26   **Figure 4.27**

Ellipse

An *ellipse* is the set of all points in the plane for which the sum of the distances from two fixed points is constant. Each of the two points is called a focus (foci are shown as F_1 and F_2 in Figure 4.27 of the ellipse. Thus, for all points P on the ellipse, the sum of the distances $(F_1P + F_2P)$ is constant. An ellipse is a flattened circle that has two centers or two foci.

The line through the two foci that connects each far edge of an ellipse is called the *major axis*. The perpendicular bisector of the major axis is called the *minor axis*. The midway point between the foci is the *center* of the ellipse. The major and minor axes meet at the center. The semimajor axis is the distance from the center through one focus to the ellipse. In Figure 4.27, the semimajor axis is labeled a. The semiminor axis is one-half of the minor axis and is labeled b.

If (x_0, y_0) are the coordinates of the center (Figure 4.27), then the equation of an ellipse with major axis of length $2a$ and minor axis of length $2b$ $(a > b)$ is

$$\frac{\left(x - x_0\right)^2}{a^2} + \frac{\left(y - y_0\right)^2}{b^2} = 1 \text{ with the major axis parallel to the } x\text{-axis.}$$

If the center of an ellipse is at the origin, the equation of the ellipse reduces to

$$\frac{x^2}{a^2} + \frac{y^2}{b^2} = 1.$$

If the major axis is parallel to the y-axis, the equation is

$$\frac{\left(x - x_0\right)^2}{b^2} + \frac{\left(y - y_0\right)^2}{a^2} = 1, \text{ or}$$

$$\frac{x^2}{b^2} + \frac{y^2}{a^2} = 1 \text{ if the center is at the origin.}$$

An ellipse is symmetrical about its major axis and its minor axis. Optical and acoustical properties of ellipses are useful in architecture and engineering. Any light or signal that starts at one focus will meet at the other focus after reflection. This property is used in wshispering galleries. You will find that many arches are portions of an ellipse because of its pleasant shape. Gears are cut into elliptic shapes that give nonuniform motion. As mentioned earlier, planets and comets move in elliptic orbits.

Note that circles and ellipses occur frequently in physics problems. These figures can be drawn by scribing a line inside a loop of string stretched out around a fixed single point (center for the circle) or two fixed points (two foci for the ellipse). Note also that a *tangent* line drawn at any point to a circle is *perpendicular* to the *radius* of the circle.

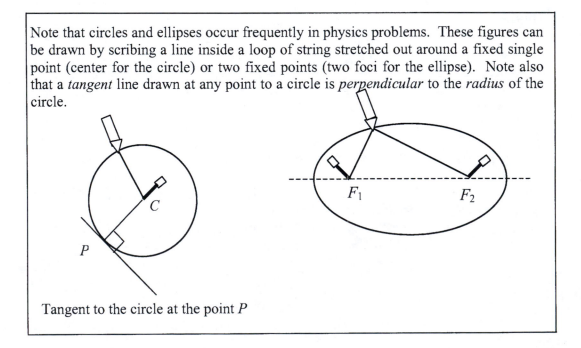

Tangent to the circle at the point P

Parabola

A *parabola* is the set of all points in the plane for which the distance from a fixed point, called the *focus*, and a fixed line, called the *directrix*, are the same. If F is the focus and PQ is the length of the perpendicular distance from a point P on the parabola on the directrix, then $FP = PQ =$ constant (Figure 4.28).

The line that passes through the focus and is perpendicular to the directrix is called the *parabolic axis*. A parabola is symmetric about its parabolic axis. The axis meets the parabola at its *vertex*.

If the vertex of a parabola is at (x_0, y_0), then the equation of a parabola that opens right or left (depending on whether a is positive or negative) is

$$x - x_0 = a \, (y - y_0)^2.$$

For a parabola that opens up or down (depending on whether a is positive or negative), the equation is

$$y - y_0 = a \, (x - x_0)^2.$$

A function of the form $y = ax^2 + bx + c$ or $x = ay^2 + by + c$ can be written in one of the preceding two forms. Thus, they represent the parabola in general.

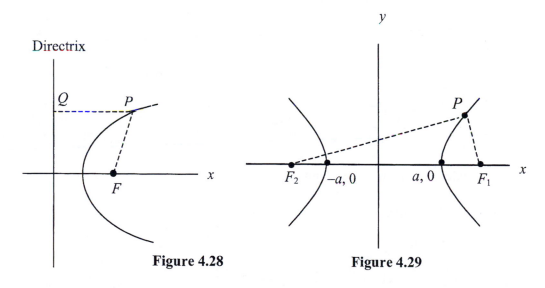

Figure 4.28   **Figure 4.29**

In architecture, you will find many arches that are parabolic in shape, because that shape provides strength. The cables that support suspension bridges hang in the form of a parabola. When a parabola is rotated about its axis, it provides a dish-shaped surface called a *paraboloid*. Paraboloids are used as flashlight and headlight reflectors since if a source of light (or sound) is placed at the focus of a paraboloid, it reflects outward in parallel paths. Parabolic surfaces are used as antennas because the surface concentrates the incoming signal at the focus. Parabolic mirrors use solar energy to concentrate light at a point to generate heat.

Hyperbola

A *hyperbola* is the set of all points for which the difference between distances from two fixed points is a constant. Each of the two points is called a *focus*. Thus, for all points P on the hyperbola whose two foci are at F_1 and F_2, $F_1P - F_2P$ = constant (Figure 4.29). Midway between the foci is the *center* of the hyperbola. The line joining the two foci is called the transverse axis (x-axis) and the perpendicular bisector of that line is called the conjugate axis (y-axis). If (x_0, y_0) are the coordinates of the center, then the equation of a hyperbola that opens left and right is

$$\frac{\left(x - x_0\right)^2}{a^2} - \frac{\left(y - y_0\right)^2}{b^2} = 1.$$

The two vertices of the hyperbola are at $(x_0 + a, y_0)$ and $(x_0 - a, y_0)$.

If (x_0, y_0) are the coordinates of the center, then the equation of a hyperbola that opens up and down is

$$\frac{\left(y - y_0\right)^2}{a^2} - \frac{\left(x - x_0\right)^2}{b^2} = 1.$$

The two vertices are at $(x_0, y_0 + a)$ and $(x_0, y_0 - a)$.

If the center is at the origin, the preceding two equations become

$$\frac{x^2}{a^2} - \frac{y^2}{b^2} = 1 \text{ and } \frac{y^2}{a^2} - \frac{x^2}{b^2} = 1, \text{respectively.}$$

A hyperbola is symmetric about its *transverse* axis (x axis) and its *conjugate* axis (y-axis). A hyperbola forms the basis of a navigation system called long-range navigation, or LORAN. By determining the difference of the distances between the ship and two radio transmitters, the LORAN system places the ship on a desired hyperbola.

There is a special type of hyperbola (centered at the origin) that does not intersect the x-axis or the y-axis. The x- and y-axes are the *asymptotes* of these hyperbolas. These hyperbolas can be expressed by the equation $xy = k$, where $k \neq 0$ and they are common in physics. For example, the relationship between the pressure P and volume V of an ideal gas when its temperature T is constant can be expressed by the equation $PV =$ constant. If V is plotted along the x-axis and its P along the y-axis, the graph is a hyperbola with the P- and V-axes as its two asymptotes (Figure 4.30).

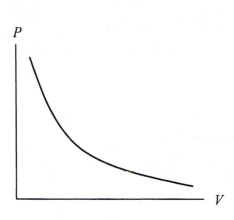

Figure 4.30

*Eccentricity of a Conic Section

The *eccentricity* ε of a circle, an ellipse, a parabola, and a hyperbola is given by

$$\varepsilon = \sqrt{1 - \left(\frac{b}{a}\right)^2}$$

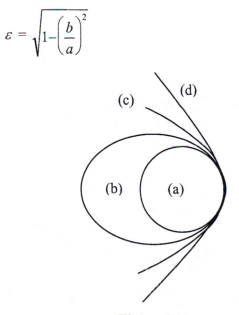

Figure 4.31

$\varepsilon = 0$ for a circle (Figure 4.31a)

$0 < \varepsilon < 1$ for an ellipse (Figure 4.*32*b)

$\varepsilon = 1$ for a parabola (Figure 4.33c)

$\varepsilon > 1$ for a hyperbola (Figure 4.34d).

For example, the eccentricity for the orbit of the moon about the earth is $\varepsilon = 0.055$. The eccentricity of the earth's orbit about the sun is $\varepsilon = 0.017$. Both orbits are, therefore, nearly circular.

Example 4.8.1
Write down the equation $x^2 - 9y^2 + 2x + 36y = 44$ in standard form, and determine what kind of conic section it represents.

$$x^2 - 9y^2 + 2x + 36y = 44$$

First rearrange the terms. $x^2 + 2x - 9y^2 + 36y = 44$

Factor out 9. $x^2 + 2x - 9(y^2 - 4y) = 44$

Complete the squares on x and y. $(x^2 + 2x + 1) - 9(y^2 - 4y + 4) = 44 + 1 - 36$

$$(x + 1)^2 - 9(y - 2)^2 = 9$$

Divide by 9 on both sides. $\dfrac{(x+1)^2}{9} - \dfrac{(y-2)^2}{1} = 1$

Comparing this equation with the standard equations of a conic section, we see that the former represents a hyperbola with its center at $(-1, 2)$.

Example 4.8.2
In the Rutherford alpha particle scattering experiment that led to the discovery of the atomic nucleus, alpha particles were shot toward a thin sheet of gold. An incoming alpha particle was repelled by the nucleus of an atom. If the nucleus was at the origin and the alpha particle traveled along a hyperbolic path whose equation was $4x^2 - y^2 = 64$, how close did the particle come to the nucleus?

The closest point of approach is the vertex of the hyperbola.

$$4x^2 - y^2 = 64$$

Divide by 64.

$$\frac{4x^2}{64} - \frac{y^2}{64} = 1$$

$$\frac{x^2}{16} - \frac{y^2}{64} = 1$$

$$\frac{x^2}{4^2} - \frac{y^2}{8^2} = 1$$

Hence, $a = 4$, and the vertex of the hyperbola is at (4, 0). The closest point of approach is then 4 units from the nucleus.

Example 4.8.3

A cannon ball follows the parabolic trajectory $y = 11x - x^2$. Here, x and y represent the horizontal and vertical distances in meters of the ball at a particular instant. Where does the ball land?

When the ball reaches the ground, $y = 0$.

Set $y = 0$ into the equation.

$$0 = 11x - x^2$$

Factor.

$$0 = x(11 - x)$$

$$x = 0, 11 \text{ m}$$

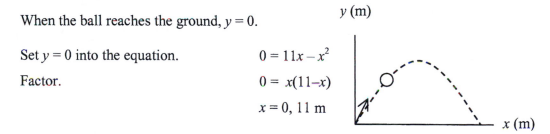

$x = 0$ m is the location from where the ball was fired. Thus, the ball lands at $x = 11$m.

Exercises

Write down the following equations in standard form and determine what kind of conic sections are represented.

7. $x^2 - 16y^2 + 10x + 9 = 0$

8. $4x^2 + 9y^2 - 16x - 18y - 11 = 0$

9. $x^2 + y^2 - 4x - 8y + 16 = 0$

10. $2x^2 - 4x - y + 4 = 0$

4.9   Parametric Equations

So far, we have described equations in which the variable y has been expressed in terms of x. There are many occasions when it is desirable to express the relationship between two variables in terms of a third variable, upon which the original two variables depend. The equations involving the third variable are called *parametric equations*. For example, the relationship between x and y for a circle passing through the origin is $x^2 + y^2 = a^2$. This relationship can be expressed in terms of a third variable t as $x = a \cos t$, $y = a \sin t$.

Parametric equations are used in kinematics, trigonometry, and calculus. The following parametric equations used in kinematics are derived in physics books for the horizontal displacement x and the vertical displacement y of an object starting from the origin and moving in a gravitational field:

$$x = x_0 + v_{x0}t,$$

$$y = y_0 + v_{y0}t - \frac{1}{2}gt^2.$$

If x_0 and $y_0 = 0$, then these equations reduce to

$$x = v_{x0}t,$$

$$y = v_{y0}t - \frac{1}{2}gt^2,$$

where v_{x0}, v_{y0}, and g are constants. The relations are used to determine the trajectory of the object. From the first equation,

$$t = \frac{x}{v_{x0}},$$

$$y = v_{y0}t - \frac{1}{2}gt^2$$

$$= v_{y0}\left(\frac{x}{v_{x0}}\right) - \frac{1}{2}g\left(\frac{x}{v_{x0}}\right)^2 = \left(\frac{v_{y0}}{v_{x0}}\right)x - \left(\frac{g}{2v_{x0}^2}\right)x^2.$$

Clearly, this equation is of the type $y = Ax + Bx^2$, so the trajectory is a parabola.

Exercises

11. Show that the equations $x = at$ and $y = b\sqrt{1 - t^2}$ parameterize an ellipse and the equations $x = at$ and $y = b\sqrt{1 + t^2}$ parameterize a hyperbola.

12. The kinematic equations involving velocity (v) and time (t), as well as distance (x) and time, are given by $v = v_0 + at$ and $x = x_0 + v_0 t + 1/2at^2$, respectively ($x_0$, v_0, and a are constants). In these equations, v and x are separately expressed in terms of t. Show that the direct relationship between v and x is given by $v^2 = v_0^2 + 2a(x - x_0)$.

4.10   Symmetry in Physics

An object or a system is said to be *symmetric* if it appears to be the same under certain operations. The operations include *translation, rotation,* and *reflection*. A circle has infinite symmetry both in translation and rotation. When you move a circle in a straight line, regardless of where you move it, the circle still has the same appearance. Also, regardless of how you rotate a circle, it appears exactly the same for an infinite number of orientations, as shown in Figure 4.32. A square has a fourfold rotational symmetry. Its view will be invariant with respect to a rotation of 90°. An equilateral triangle has a threefold rotational symmetry, since it looks the same from any one of the three corners—that is, with respect to a rotation of 120°. A parabola has a *mirror symmetry*, because if you place a mirror perpendicularly on its parabolic axis, you will see its other half in the mirror.

Knowledge of symmetry is important in physics. Take advantage of symmetry when you solve physics problems. For example, there is a symmetry for the motion of an object thrown vertically upward with an initial speed. The object slows down at a constant rate as it moves up, stops momentarily at its maximum height, and then speeds up at the same rate as it falls down. *Time symmetry* exists for this motion, in the sense that the time required for the object to reach its maximum height is equal to the time for its return to its starting point. There is also a *symmetry in the speed* of the object, in the sense that at any distance above the point of release, the speed of the object during the upward trip is equal to the speed at the same point during the downward trip (Figure 4.33a). A similar symmetry exists for the motion of a projectile that has been shot at an angle other than 90° or 0°. In this case, the trajectory is a parabola, but the time to reach the maximum height is one-half of the total time of flight, and at any height, the vertical components of velocity are equal and opposite (Figure 4.33b).

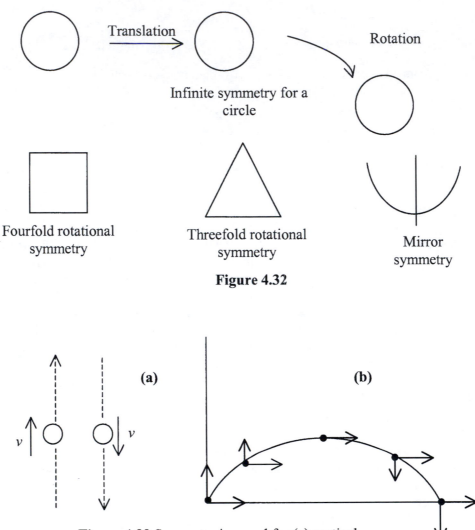

Figure 4.32

Figure 4.33 Symmetry in speed for (a) vertical motion and (b) projectile motion.

The universe has different symmetries. It is interesting and important to know that the symmetries of the universe are related to the fundamental conservation principles of physics (conservation of linear momentum, conservation of angular momentum, and conservation of energy). The conservation of linear momentum implies that the laws of physics have a *translational* symmetry or, in other words, the universe exhibits *homogeneity of space*. The conservation of angular momentum implies that the laws of physics have a *rotational symmetry* or, in other words, the universe exhibits *isotropy of space*. The conservation of energy implies that laws of physics do not change with time, that is, have a *temporal symmetry* or, in other words, the universe exhibits *homogeneity of time*. You will learn about these conservation principles from your physics course. In fact, *for any conservation principle, there exists a type of symmetry in nature and vice versa*. This is known as *Noether's theorem*.

*4.11 Non-Euclidean Geometry

The geometry that's been discussed so far is called *plane geometry* or *Euclidean geometry*. In this geometry, the shortest distance between two points is a straight line, and the sum of the three angles in a triangle is 180°. But suppose you need to know the shortest flight path between two airports that are very far apart.

Since the earth is nearly a sphere, this flight path would be different from the one gotten by using a ruler to draw a straight line on a flat map joining the two airports on paper.

The geometry on a flat surface is Euclid's; geometry on a curved surface (such as the surface of the earth) is non-Euclidean. To determine the path of an aircraft or the distance between two cities, non-Euclidean geometry is relevant.

In the language of geometry, the two-dimensional surface of a sphere (such as the earth) is curved, and the surface has *positive curvature*. The sum of three angles of a triangle drawn connecting three points on the surface of the sphere is *more* than 180°. For example, start at the North Pole and move to the equator. Turn 90° and move along the equator a quarter of the way around the earth. Then, turn 90° to go to the pole. You will end up 90° from your original path. The triangle has three angles of 90° each. So the sum of the angles is 270°.

The two-dimensional surface on a flat piece of paper has *zero curvature*, and the sum of the three angles of a triangle on the paper is 180°. A saddle-shaped surface is also curved, but it has *negative curvature*.

The sum of the three angles of a triangle on a surface of negative curvature is *less* than 180°. Figure 4.34 shows surfaces with zero, positive, and negative curvatures and also shows what a triangle drawn on these surfaces looks. On a curved surface, the lines forming straight lines are not all straight from a three-dimensional point of view, but they are the shortest distances between points if we are confined to the curved surface.

Zero curvature   Positive curvature   Negative curvature

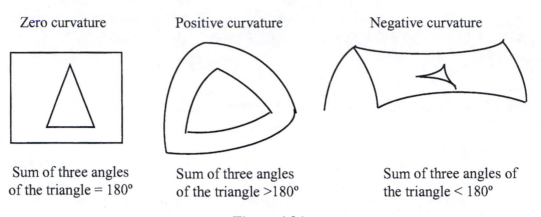

Sum of three angles   Sum of three angles   Sum of three angles of
of the triangle = 180°   of the triangle >180°   the triangle < 180°

Figure 4.34

Non-Euclidean geometry is useful when a surface is curved. For example, in the general theory of relativity, which is a new interpretation of gravitation, gravity is related to the *geometry of space–time*, and space–time is *curved* when gravity is strong. Non-Euclidean geometry is useful in studying the general theory of relativity. Because space–time is curved near the sun owing to its own gravity, light beams bend when passing near the sun. You will learn more about this phenomenon from your introductory physics course.

Answers to Exercises in Chapter 4

1. 3.0×10^{-4} rad, 4 cm

2. $A = 30°$, $B = 115°$, $C = 55°$, $D = 60°$

3. $\psi = 75°$, $\phi = 75°$, $\eta = 60°$, $RS = 2.8$, $RT = 3.2$, $\theta = 45°$

4. 5490 kg/m^3

5. 0.2 cm^2

6. 7.3 cm^3

7. $(x + 5)^2/16 - y^2 = 1$, hyperbola

8. $(x - 2)^2/9 + (y - 1)^2/4 = 1$, ellipse

9. $(x - 2)^2 + (y - 4)^2 = 4$, circle

10. $y = 2(x - 1)^2 + 2$, parabola

Chapter 5
Trigonometry

Trigonometry deals with the interrelationships among the sides and angles of triangles. It is based primarily on the special properties of a right triangle. It is a powerful tool for scientists and engineers for taking measurements as well as for solving real-world problems.

5.1   Angles in Trigonometry

In trigonometry, *angles* are created because of the *rotation*, in a plane, of a straight line about a point O on the line, called the *origin*.   If the line OA is rotated in a plane about O onto OB, an angle AOB is generated (Figure 5.1). The angle is considered to be *positive* or *negative* depending on whether the rotation is *counterclockwise* or *clockwise*, respectively. As mentioned in Chapter 4, angles can be measured in degrees or in radians.

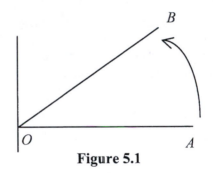

Figure 5.1

$$360° = 2\pi \text{ radians}, \qquad 1 \text{ radian} = \frac{180}{\pi} \text{ degrees} = 57.3° \qquad 1° = \frac{\pi}{180} \text{ radians}$$

An angle is considered to be in a standard position if its vertex is at the origin and its initial side is on the positive *x*-axis of a rectangular Cartesian coordinate system. An angle is considered to be a first-, second-, third-, or fourth-quadrant angle, depending on whether its terminal side falls in the first, second, third, or fourth quadrant, respectively, of the coordinate system (Figure 5.2).

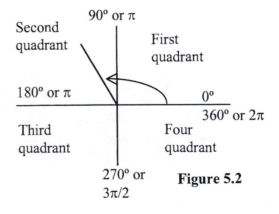

Figure 5.2

Right-Angle Trigonometry

5.2   Right Triangles and the Pythagorean Theorem

A *right triangle* is a triangle in which one angle is 90°. The lengths of the base b, height a, and hypotenuse c, of a right triangle (Figure 5.3) are related by

$$c^2 = a^2 + b^2.$$

This is known as the *Pythogorean theorem*, a mathematical theorem that is very useful for finding the length of one side of a right triangle when the other two sides are known.

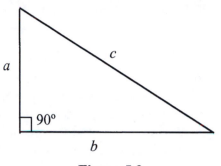

Figure 5.3

In physics, the Pythogorean theorem is useful in many practical situations. An example in kinematics is to determine the displacement of an object, as illustrated in the following example.

Example 5.2.1

The amount of displacement of an object is defined as the shortest distance between its final location and its initial location. A person travels 3 mi east and then travels 4 mi north. What is the amount of displacement of the person from his initial location?

The person has traveled along the two perpendicular sides of a right triangle. The sketch of the problem is shown on the right. The hypotenuse of the triangle is the required displacement.
Thus, in this case,
  $b = 3$ mi and $a = 4$ mi.
From the Pythagorean theorem, the magnitude of the displacement is given by
$$c^2 = a^2 + b^2$$
$$= (3 \text{ mi})^2 + (4\text{mi})^2$$
$$= 9 \text{ mi}^2 + 16 \text{ mi}^2 = 25 \text{ mi}^2.$$

The displacement is, then, $c = \sqrt{25}$ mi $= 5$ mi.
The result makes sense because the hypotenuse of a right triangle is always greater than the other two sides.

5.3   Trigonometric Functions

Three *trigonometric functions* are often used in physics. These are defined in terms of the ratio of two sides of a right triangle. The functions are called *sine* (or *sin*), *cosine* (or *cos*), and *tangent* (or *tan*). Consider a right triangle whose one acute angle is θ at one vertex as shown in Figure 5.4. Its *hypotenuse* is h, the side *opposite* θ is h_o (also called *rise*), and the side *adjacent* to θ is h_a (also called *run*.) Then *sin*, *cos*, and *tan* are defined by the following ratios:

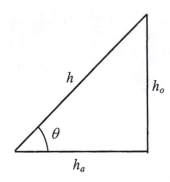

Figure 5.4

$$\sin \theta = \frac{h_o}{h} = \frac{\text{opposite}}{\text{hypotenuse}} \text{ or } \frac{\text{rise}}{\text{hypotenuse}}$$

$$= \frac{h_o}{\sqrt{h_o^2 + h_a^2}} \text{ (Since from the Pythagorean theorem } h^2 = h_a^2 + h_o^2)$$

$$\cos \theta = \frac{h_a}{h} = \frac{\text{adjacent}}{\text{hypotenuse}} \text{ or } \frac{\text{run}}{\text{hypotenuse}} = \frac{h_a}{\sqrt{h_o^2 + h_a^2}}$$

$$\tan \theta = \frac{h_o}{h_a} = \frac{\text{opposite}}{\text{adjacent}} \text{ or } \frac{\text{rise}}{\text{run}}.$$

The other three trigonometric functions are *cosecant* (or *csc*), *secant* (or *sec*), and *cotangent* (or *cot*). These are defined as follows:

$$\csc \theta = \frac{h}{h_o} = \frac{1}{\sin \theta}$$

$$\sec \theta = \frac{h}{h_a} = \frac{1}{\cos \theta}$$

$$\cot \theta = \frac{h_a}{h_o} = \frac{1}{\tan \theta}.$$

Now, $\tan \theta = \dfrac{h_o}{h_a}$. This can be written as

$$\tan \theta = \frac{h_o}{h_a} = \frac{h_o/h}{h_a/h} \quad \text{(by dividing the numerator and denominator by } h\text{).}$$

From the definitions of *sin* and *cos*, it is seen that the right-hand side of the preceding equation is the ratio of *sin* and *cos*. Therefore,

$$\tan \theta = \frac{\sin \theta}{\cos \theta}.$$

Similarly, we can show that $\cot \theta = \dfrac{\cos \theta}{\sin \theta}$.

In most situations in physics, you need to know only sin, cos, and tan. An easy way to remember how they are defined is to remember the mnemonic *soh cah toa* (usually pronounced as "so-kah-toe-ah."). This means that

$$\mathbf{soh} = \sin = \frac{\text{opposite}}{\text{hypotenuse}}.$$

$$\mathbf{cah} = \cos = \frac{\text{adjacent}}{\text{hypotenuse}}.$$

$$\mathbf{toa} = \tan = \frac{\text{opposite}}{\text{adjacent}}.$$

The values of sin, cos, and tan for commonly used angles are shown in Table 5.1.

Table 5.1 Values of sin, cos, and tan for Commonly Used Angles

θ	$\sin \theta$	$\cos \theta$	$\tan \theta$
0°	0	1	0
30°	$\frac{1}{2} = 0.5$	$\sqrt{3}/2 = 0.87$	$1/\sqrt{3} = 0.58$
45°	$1/\sqrt{2} = 0.71$	$1/\sqrt{2} = 0.71$	1
60°	$\sqrt{3}/2 = 0.87$	$\frac{1}{2} = 0.5$	$\sqrt{3} = 1.73$
90°	1	1	∞

Figure 5.5 shows the signs (+ or –) that sin, cos, and tan take on for angle θ in different quadrants (0 to 2π or 360°). The angles are measured *counterclockwise* from the *x*-axis, as shown in the figure. The signs are determined by the sign of *x* and *y* in different quadrants. The negative angles are measured below the *x*-axis. This means that –60° is 360° – 60° = 300°.

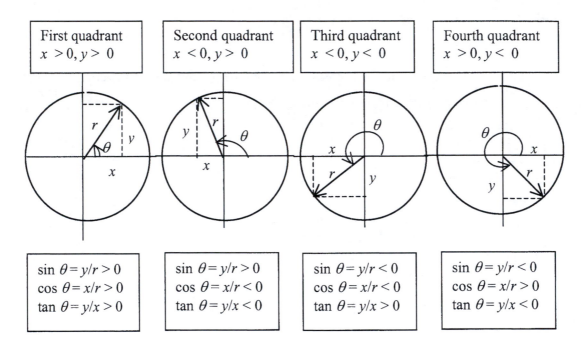

First quadrant $x > 0, y > 0$	Second quadrant $x < 0, y > 0$	Third quadrant $x < 0, y < 0$	Fourth quadrant $x > 0, y < 0$
$\sin \theta = y/r > 0$ $\cos \theta = x/r > 0$ $\tan \theta = y/x > 0$	$\sin \theta = y/r > 0$ $\cos \theta = x/r < 0$ $\tan \theta = y/x < 0$	$\sin \theta = y/r < 0$ $\cos \theta = x/r < 0$ $\tan \theta = y/x > 0$	$\sin \theta = y/r < 0$ $\cos \theta = x/r > 0$ $\tan \theta = y/x < 0$

Figure 5.5

This is usually remembered as *all, sin, tan, cos*, meaning *all* are positive in the first quadrant, *sin* is positive in the second quadrant, *tan* is positive in the third quadrant, and *cos* is positive in the fourth quadrant.

The signs of sin, cos, and tan in four quadrants are also shown in Table 5.2.

The plots of the sin, cos, and tan for -2π to $+2\pi$ are shown in Figure 5.6.

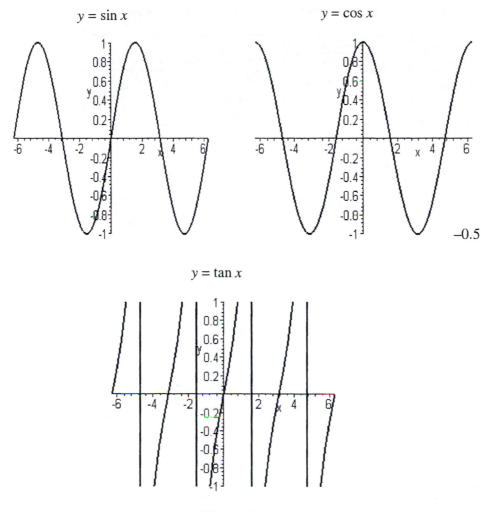

$y = \sin x$

$y = \cos x$

-0.5

$y = \tan x$

Figure 5.6

Table 5.2 Sign of Trigonometric Functions in Four Quadrants

Quadrant	sin	cos	tan
I	+	+	+
II	+	−	−
III	−	−	+
IV	−	+	−

The plots also show that the signs of the trigonometric functions can be positive and negative. The variations of the values of sin, cos, and tan, as seen from their plots, are summarized in Table 5.3.

Table 5.3 Variations of sin, cos, and tan in Four Quadrants

Quadrant	sin	cos	tan
I	0 to +1	+1 to 0	0 to +∞
II	+1 to 0	0 to −1	−∞ to 0
III	0 to −1	−1 to 0	0 to +∞
IV	−1 to 0	0 to +1	−∞ to 0

Note: In general,

$$-1 \leq \sin \theta \leq +1$$

$$-1 \leq \cos \theta \leq +1$$

$$-\infty \leq \tan \theta \leq \infty$$

Example 5.3.1

Find the height of an equilateral triangle whose sides are each 10 cm in length.

We can solve the problem by using (i) the Pythagorean theorem, as well as (ii) a trigonometric function.

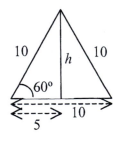

(i) The height h bisects the base. Therefore, from the Pythagorean theorem,

$$5^2 + h^2 = 10^2$$

or $h = \sqrt{(10\,\text{cm})^2 - (5\text{cm})^2} = 8.7$ cm.

(ii) For an equilateral triangle, all angles are 60°. Therefore, using the *sin* function, we obtain

$$\sin 60° = \frac{h}{10\,\text{cm}} \quad [\text{since } \sin \theta = \frac{\text{opposite}}{\text{hypotenuse}}].$$

Multiply by 10. $h = (10\,\text{cm}) \sin 60° = (10\,\text{cm}) \times 0.866 = 8.7$ cm

Example 5.3.2

On a sunny day, a tall pillar casts a shadow that is 6.3 m long. The angle between the sun's rays and the ground is 47°. Determine the height of the pillar.

For this problem, the angle $\theta = 47°$ and the distance (h_a or the "adjacent") = 6.3 m.

The height (h_o or "opposite") is unknown.

The trigonometric function that

involves θ, h_a, and h_o is tan θ.

$$\tan \theta \quad = \frac{h_o}{h_a} \text{ or } \frac{\text{opposite}}{\text{adjacent}}$$

or $h_o = h_a \tan \theta$

Substitute the values.   = 6.3 m X tan (47°)

= 6.3 m X 1.07

= 6.5 m

Therefore, the height of the pillar is 6.5 m.

Example 5.3.3

A box is pulled up a 36° incline. When it has moved 4.0 m along the incline, what distance x has it moved in the horizontal direction and what elevation y has it achieved?

In this case, $\theta = 36°$, h or hypotenuse = 4.0 m

h_o or "opposite" = y (to be determined)

h_a or "adjacent" = x (to be determined)

$$\sin \theta = \frac{h_o}{h} \text{ or } \frac{\text{opposite}}{\text{hypotenuse}}$$

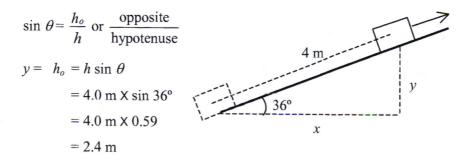

$y = h_o = h \sin \theta$

= 4.0 m X sin 36°

= 4.0 m X 0.59

= 2.4 m

$$\cos\theta = \frac{h_a}{h} \quad \text{or} \quad \frac{\text{adjacent}}{\text{hypotenuse}}$$

$$x = h_a = h \cos\theta = 4.0 \text{ m} \times \cos 36° = 4.0 \text{ m} \times 0.81 = 3.2 \text{ m}$$

Exercises

Using trigonometric functions, determine the lengths of the sides of the right triangles that are not given, using trigonometric functions.

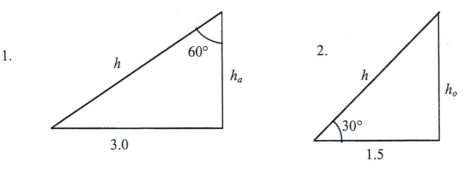

1.

2.

3. In a stairway, each step is set back 25 cm from the next lower step and the stairway rises at an angle of 38° with respect to the horizontal. Determine the height of each step.

4. Where the road is curved, the roadbed is elevated at an angle with respect to the horizontal to avoid skidding when a car follows the curved road. This angle is called a banking angle. If the banking angle is 15° and the outside of the curved road is elevated by 2.0 m from the inside of the curve, determine the width of the road where it is curved.

5. The gondola ski lift in Colorado is 2830 m long. If the ski lift rises 14.6° from the horizontal, determine the height of the top of the lift relative to the ground.

5.4   Common Trigonometric Identities

From the Pythagorean theorem and using Figure 5.4, we find that

$h^2 = h_a^2 + h_o^2 = (h \sin \theta)^2 + (h \cos \theta)^2$ (from the definitions of *sin* and *cos*).

Canceling h^2 from both sides, we get

$\sin^2 \theta + \cos^2 \theta = 1$.

This is a *trigonometric identity*. There are several other trigonometric identities, most of which can easily be proved.

The following trigonometric identities and relations are useful in physics and in solving problems:

Pythagorean Relations
$\sin^2 \theta + \cos^2 \theta = 1$ $\sec^2 \theta = 1 + \tan^2 \theta$

$\csc^2 \theta = 1 + \cot^2 \theta$

Reciprocal Relations

$\csc \theta = \dfrac{1}{\sin \theta}$ $\sec \theta = \dfrac{1}{\cos \theta}$

$\cot \theta = \dfrac{1}{\tan \theta}$

Quotient Relations

$\tan \theta = \dfrac{\sin \theta}{\cos \theta}$ $\cot \theta = \dfrac{\cos \theta}{\sin \theta}$

Example 5.4.1

Show that $\sec\theta - \cos\theta = \tan\theta \cdot \sin\theta$.

We have	$\sec\theta - \cos\theta$.
Substitute $1/\cos\theta$ for $\sec\theta$.	$= 1/\cos\theta - \cos\theta$
	$= \dfrac{1 - \cos^2\theta}{\cos\theta}$
Substitute $\sin^2\theta$ for $1 - \cos^2\theta$.	$= \dfrac{\sin^2\theta}{\cos\theta} \qquad (\sin^2\theta + \cos^2\theta = 1)$
	$= \dfrac{\sin\theta}{\cos\theta} \cdot \sin\theta$.
Substitute $\tan\theta$ for $\sin\theta/\cos\theta$.	$= \tan\theta \cdot \sin\theta$

Exercises

Prove the following:

6. $\dfrac{\sin\theta}{1 + \cos\theta} = \dfrac{1 - \cos\theta}{\sin\theta}$ [*Hint*: Multiply the left-hand side by $(1 - \cos\theta)$.]

7. $\dfrac{1 - \tan^2\theta}{1 + \tan^2\theta} = 1 - 2\sin^2\theta$ [*Hint*: $\tan\theta = \dfrac{\sin\theta}{\cos\theta}$.]

8. $\dfrac{\tan\theta - \cot\theta}{\tan\theta + \cot\theta} = 1 - 2\cos^2\theta$ 9. $\dfrac{\sec\theta}{\tan\theta + \cot\theta} = \sin\theta$

Angle-Sum and Angle-Difference Relations

$\sin(A + B) = \sin A \cos B + \cos A \sin B$

$\sin(A - B) = \sin A \cos B - \cos A \sin B$

$\cos(A + B) = \cos A \cos B - \sin A \sin B$

$\cos(A - B) = \cos A \cos B + \sin A \sin B$

$\tan(A + B) = \dfrac{\tan A + \tan B}{1 - \tan A \tan B}$ $\tan(A - B) = \dfrac{\tan A - \tan B}{1 + \tan A \tan B}$

$\cot(A + B) = \dfrac{\cot A \cot B - 1}{\cot B + \cot A}$ $\cot(A - B) = \dfrac{\cot A \cot B + 1}{\cot B - \cot A}$

Example 5.4.2
Show that $\sin 2A = 2\sin A \cos A$.

We have $\qquad\qquad\qquad\qquad\qquad\qquad \sin 2A = \sin(A + A).$

Use the formula for $\sin(A + B)$. $\qquad\qquad = \sin A \cos A + \cos A \sin A$

$\qquad\qquad\qquad\qquad\qquad\qquad\qquad = 2\sin A \cos A.$

Exercises
Show that the following are true:

10. $\tan(\theta + \pi/4) = \dfrac{\cos\theta + \sin\theta}{\cos\theta - \sin\theta}$ $\qquad$ 11. $\sin(A + B)\sin(A - B) = \sin^2 A - \sin^2 B$

12. $\cos(\pi/3 + B) = \sin(\pi/6 - B)$

13. $\cos(A + B)\cos B + \sin(A + B)\sin B = \cos A$

Double-Angle Relations

$$\sin 2\theta = 2\sin\theta\cos\theta = \frac{2\tan\theta}{1 + \tan^2\theta}$$

$$\cos 2\theta = \cos^2\theta - \sin^2\theta = 2\cos^2\theta - 1 = 1 - 2\sin^2\theta = \frac{1 - \tan^2\theta}{1 + \tan^2\theta}$$

$$\tan 2\theta = \frac{2\tan\theta}{1 - \tan^2\theta} \qquad\qquad \cot 2\theta = \frac{\cot^2\theta - 1}{2\cot\theta}$$

Example 5.4.3
Show that $\cos^4 A - \sin^4 A = \cos 2A$.

We have $\qquad\qquad\qquad \cos^4 A - \sin^4 A = (\cos^2 A)^2 - (\sin^2 A)^2$

$\qquad\qquad\qquad\qquad\qquad\qquad\qquad = (\cos^2 A + \sin^2 A)(\cos^2 A - \sin^2 A)$

Substitute 1 for $\cos^2 A + \sin^2 A$. $\qquad = 1 \cdot (\cos^2 A - \sin^2 A)$

$\qquad\qquad\qquad\qquad\qquad\qquad\qquad = \cos^2 A - \sin^2 A = \cos 2A.$

Exercises
Show that the following identities are true:

14. $\dfrac{1 + \sin 2\theta}{1 + \cos 2\theta} = \frac{1}{2}(1 + \tan\theta)^2$ $\qquad$ 15. $\dfrac{1 - \cos 2\theta}{1 + \cos 2\theta} = \tan^2\theta$

16. $\cos 2\theta = \dfrac{1 - \tan^2\theta}{1 + \tan^2\theta}$

17. $\dfrac{\sin 2\theta}{\sin\theta} + \dfrac{\cos 2\theta}{\cos\theta} = \dfrac{2\sin 3\theta}{\sin 2\theta}$

Function-Sum and Function-Difference Relations

$$\sin A + \sin B = 2\sin\left(\frac{A+B}{2}\right)\cos\left(\frac{A-B}{2}\right)$$

$$\sin A - \sin B = 2\sin\left(\frac{A-B}{2}\right)\cos\left(\frac{A+B}{2}\right)$$

$$\cos A + \cos B = 2\cos\left(\frac{A+B}{2}\right)\cos\left(\frac{A-B}{2}\right)$$

$$\cos A - \cos B = 2\sin\left(\frac{A+B}{2}\right)\sin\left(\frac{B-A}{2}\right)$$

The following example shows how the function-sum relation is used in describing a physical phenomenon.

Example 5.4.4 (Standing wave)
Two identical waves traveling in opposite directions can be represented by the equations $y_1 = A\sin(kx - \omega t)$ and $y_2 = A\sin(kx + \omega t)$, where A, k, and ω are characteristic constants of the waves (called amplitude, wave number, and angular frequency, respectively) and x and t represent position and time, respectively.   Show that the resulting wave $y = y_1 + y_2 = 2A\sin kx\cos\omega t$.

Following is a sketch of the problem:

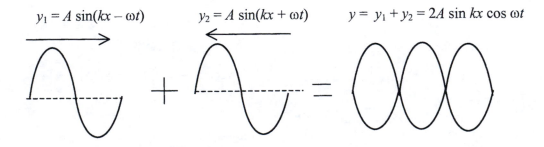

$y_1 = A\sin(kx - \omega t)$ $y_2 = A\sin(kx + \omega t)$ $y = y_1 + y_2 = 2A\sin kx\cos\omega t$

To solve the problem mathematically, use the addition relation for sine functions:

$$y = y_1 + y_2$$

$$= A \sin(kx - \omega t) + A \sin(kx + \omega t)$$

Factor out A. $= A[\sin(kx - \omega t) + \sin(kx + \omega t)]$

Use the addition relation. $= A \cdot 2 \sin \dfrac{(kx - \omega t) + (kx + \omega t)}{2} \cos \dfrac{(kx - \omega t) - (kx + \omega t)}{2}$

(since $\sin A + \sin B = 2 \sin (A + B)/2 \cos (A - B)/2$)

$$= 2A \sin\left(\dfrac{2kx}{2}\right) \cos\left(\dfrac{-2\omega t}{2}\right).$$

Cancel 2 from *sin* and *cos* $= 2A \sin kx \cos (-\omega t)$

$= 2A \sin kx \cos \omega t.$ (since $\cos (-\theta) = \cos \theta$, see the *reduction formulas* that follow.)

The resulting wave y is known as a standing wave in physics.

Exercises

18. Two waves traveling in a medium are represented by the equations $y_1 = A \sin (kx - \omega t)$ and $y_2 = A \sin (kx - \omega t + \phi)$ where A, k, and ω are characteristic constants of the waves (called amplitude, wave number, and angular frequency, respectively), ϕ is called the phase difference between the two waves, and x and t represent position and time, respectively.   Show that the resulting wave

$$y = y_1 + y_2 = 2A \cos \dfrac{\phi}{2} \sin\left(kx - \omega t + \dfrac{\phi}{2}\right).$$

(The resulting wave shows interference of waves.)

19. Two sinusoidal waves of frequencies f_1 and f_2 are represented at a fixed point in space by $y_1 = \sin 2\pi f_1 t$ and $y_2 = \sin 2\pi f_2 t$.   Show that the resultant displacement

$$y = y_1 + y_2 = 2A \cos 2\pi\left(\dfrac{f_1 - f_2}{2}\right)t \sin 2\pi\left(\dfrac{f_1 + f_2}{2}\right)t.$$

(The resulting wave shows what is called beats.)

Function-Product Relations

$$\sin A \cos B = \frac{1}{2}\sin(A+B) + \frac{1}{2}\sin(A-B)$$

$$\cos A \cos B = \frac{1}{2}\cos(A-B) + \frac{1}{2}\cos(A+B)$$

$$\sin A \sin B = \frac{1}{2}\cos(A-B) - \frac{1}{2}\cos(A+B)$$

$$\cos A \sin B = \frac{1}{2}\sin(A+B) - \frac{1}{2}\sin(A-B)$$

Half -Angle Relations

$$\sin \frac{\theta}{2} = \sqrt{\frac{1-\cos\theta}{2}} \qquad \tan \frac{\theta}{2} = \sqrt{\frac{1-\cos\theta}{1+\cos\theta}}$$

$$\cos \frac{\theta}{2} = \sqrt{\frac{1+\cos\theta}{2}} \qquad \cot \frac{\theta}{2} = \sqrt{\frac{1+\cos\theta}{1-\cos\theta}}$$

Exercises

Show that the following identities are true:

20. $\cot \dfrac{\theta}{2} = \dfrac{1}{\csc\theta - \cot\theta}$

21. $\tan \dfrac{\theta}{2} = \dfrac{\sin\theta}{1+\cos\theta}$

Reduction Formulas

$$\cos(-\theta) = +\cos\theta \qquad\qquad \sin(270° \pm \theta) = -\cos\theta$$
$$\sin(-\theta) = -\sin\theta \qquad\qquad \cos(270° \pm \theta) = \pm\sin\theta$$
$$\tan(-\theta) = -\tan\theta \qquad\qquad \tan(270° \pm \theta) = \mp\cot\theta$$

$$\sin(90° \pm \theta) = +\cos\theta \qquad\qquad \sin(360° \pm \theta) = \pm\sin\theta$$
$$\cos(90° \pm \theta) = \mp\sin\theta \qquad\qquad \cos(360° \pm \theta) = +\cos\theta$$
$$\tan(90° \pm \theta) = \mp\cot\theta \qquad\qquad \tan(360° \pm \theta) = \pm\tan\theta$$

$$\sin(180° \pm \theta) = \mp\sin\theta$$
$$\cos(180° \pm \theta) = -\cos\theta$$
$$\tan(180° \pm \theta) = \pm\tan\theta$$

Example 5.4.5

Express the following trigonometric functions in terms of angles in the first quadrant: $\sin(-60°)$, $\sin 150°$, $\cos(-60°)$, $\cos 330°$, $\tan 120°$, $\tan 210°$.
Then use Table 5.1 for the answer. Do not use a calculator.

$$\sin(-60°) \ = -\sin(60°) \ (\text{since } \sin(-\theta) = \ -\sin\theta)$$

$$= -\,0.87$$

$$\sin 150° = \sin(180 - 30)° \ = \sin 30° \ (\text{since } \sin(180° - \theta) = \ \sin\theta)$$

$$= 0.5$$

$$\cos(-60°) = \cos 60° \ (\text{since } \cos(-\theta) = \ +\cos\,\theta)$$

$$= 0.5$$

$$\tan 120° = \tan(180 - 60)° = -\tan 60° \ \ (\text{since } \tan(180° - \theta) = -\tan\,\theta)$$
$$= -1.73$$

$$\tan 210° = \tan(180 + 30)° = \ \tan 30° \ \ (\text{since } \tan(180° + \theta) = \ \tan\,\theta)$$
$$= 0.58$$

5.5   Inverse Trigonometric Functions

Any angle θ whose sin is x is denoted by arcsin x or $\sin^{-1}x$. Thus, if $\sin\,\theta = x$, $\theta = \sin^{-1}x$. Similar notation is used to define other inverse trigonometric functions: $\cos^{-1}x$, $\tan^{-1}x$, $\csc^{-1}x$, $\sec^{-1}x$, and $\cot^{-1}x$. The plots of $\sin^{-1}x$, $\cos^{-1}x$, and $\tan^{-1}x$ are shown in Figure 5.7.

> **Note:** The notation $sin^{-1}x$ is not the same as $(\sin x)^{-1}$.
>
> $(\sin x)^{-1} = 1/\sin x = \csc x$. But $\sin^{-1}x$ is the inverse sin x function.

If any inverse trigonometric function, such as $\sin^{-1}x$, is known, it is possible to express all trigonometric functions in terms of x. For example, let $\theta = \sin^{-1}x$. Therefore, $\sin\,\theta = x$. You can determine other trigonometric functions in terms of x, as the following shows:

$$\cos\,\theta \ \ = \sqrt{1 - \sin^2\theta} \qquad\qquad (\text{since } \sin^2\theta + \cos^2\theta = 1)$$

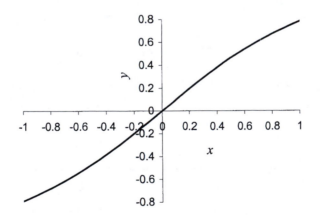

Figure 5.7

$$= \sqrt{1-x^2} \qquad\qquad \tan\theta = \frac{\sin\theta}{\cos\theta} = \frac{x}{\sqrt{1-x^2}}$$

$$\csc\theta = \frac{1}{\sin\theta} = \frac{1}{x} \qquad\qquad \sec\theta = = \frac{1}{\cos\theta} = \frac{1}{\sqrt{1-x^2}}$$

$$\cot \theta = = \frac{1}{\tan \theta} = \frac{\sqrt{1-x^2}}{x}.$$

Two useful fundamental identities involving inverse trigonometric functions are

$$\sin^{-1}x + \cos^{-1}x = \pi/2$$

and

$$\tan^{-1}x + \cot^{-1}x = \pi/2.$$

Note: Inverse trigonometric functions can be determined easily by pressing $\sin^{-1}$, $\cos^{-1}$, and $\tan^{-1}$ on your calculator. When you use a calculator to determine the value, note that the calculator always displays the smallest correct angle. For example, $\sin^{-1}(0.5)$ is both 30° and 150°, but calculator will display only 30°. Similarly, $\sin^{-1}(-0.5)$ is –30° and 210°, but the calculator will display only –30°. The displayed value *may not* be the right answer in some situations. Use the knowledge of which quadrant you are working in to find which answer is correct.

Example 5.5.1

A lakefront drops off gradually at an angle θ, as shown. A lifeguard rows out straight 20.0 m and drops a weighted fishing line to find the depth at that distance. He found the depth to be 1.28 m. What is the angle θ?

We know that

$$\tan \theta \;=\; \frac{h_o}{h_a} \;\text{ or }\; \frac{\text{opposite}}{\text{adjacent}}.$$

Substitute the values. $= \dfrac{1.28\,\text{m}}{20.0\,\text{m}}$

Divide the numbers. $= 0.064.$

Take the *inverse tan* to get the result. $\theta = \tan^{-1}(0.064) = 3.7°$

Exercises

22. Find the values of the following operations:

$$\cos^{-1}(0.56), \quad \tan^{-1}(2.5), \quad \text{and} \quad \sin^{-1}(0.92).$$

Using inverse trigonometric functions, determine the angle θ of the right triangles:
23.

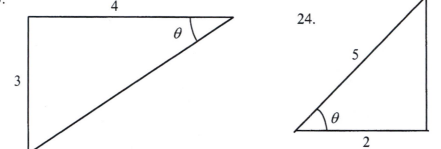

25. If $\theta = \cos^{-1}x$, determine $\sin\theta$, $\cos\theta$, $\tan\theta$, $\cot\theta$, $\csc\theta$, and $\sec\theta$ in terms of x.

26. If $\theta = \tan^{-1}x$, determine $\sin\theta$, $\cos\theta$, $\tan\theta$, $\cot\theta$, $\csc\theta$, and $\sec\theta$ in terms of x.

5.6 Series Expansion of Trigonometric Functions and Small-Angle Approximations

Trigonometric functions and inverse trigonometric functions can be expanded as follows:

$$\sin x = x - \frac{x^3}{3!} + \frac{x^5}{5!} - \ldots \qquad\qquad \sin^{-1} x = x + \frac{1}{6}x^3 + \frac{3}{40}x^5 + \ldots$$

$$\cos x = 1 - \frac{x^2}{2!} + \frac{x^4}{4!} - \ldots \qquad\qquad \cos^{-1} x = \frac{\pi}{2} - \sin^{-1} x$$

$$\tan x = x + \frac{x^3}{3} + \frac{x^5}{15} + \ldots \qquad\qquad \tan^{-1} x = x - \frac{x^3}{3} + \frac{x^5}{5} - \ldots \quad .$$

The right-hand side of the preceding expressions contains an infinite number of terms. A polynomial with an infinite number of terms is called a *power series*; therefore, the preceding expansions are called power series expansions of trigonometric or inverse trigonometric functions.

It is often desirable to *approximate* the solution of a problem because the exact solution may be difficult to obtain or may not be necessary. The preceding power series expansions are very helpful for approximations. For example, when x is in *radians* and is very small ($x \ll 1$), then x^2 and other terms are negligibly small and

$$\sin x \approx x$$

$$\cos x \approx 1 - \frac{x^2}{2} \qquad \text{(keeping up to the second power of } x\text{)}$$

$$\approx 1 \quad \text{(keeping only the first term)}$$

$$\tan x \approx x, \text{ etc.}$$

Example 5.6.1

Determine the small-angle approximation for the expression $4 \sin^2 \theta / \cos \theta$.

$$\frac{4 \sin^2 \theta}{\cos \theta} \approx \frac{4(\theta)^2}{1} \qquad \text{(since } \sin \theta \approx \theta \text{ and } \cos \theta \approx 1\text{)}$$

$$= 4\theta^2$$

Exercise

27. A small mass attached to a string attached to a fixed vertical support forms a simple pendulum. When the mass is moved to one side from its position of rest by an angle, say, θ, with the vertical line and then released, it moves back and forth. The time the mass takes to complete one full swing is called its period T and is given by

$$T = 2\pi \sqrt{\frac{L}{g}} \, [1 + \tfrac{1}{4} \sin^2 (\theta/2) + 9/64 \sin^4 (\theta/2) + \ldots],$$

where L is the length of the pendulum (the distance of the string from the fixed support plus the radius of the mass, assuming the mass is a uniform sphere), and g is the local acceleration due to gravity.
Show that for small θ, the period

can be expressed as $T = 2\pi \sqrt{\dfrac{L}{g}} \, (1 + \theta^2/16)$.

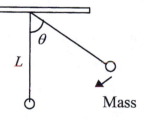

Non-Right-Angle Trigonometry

5.7   Laws of Sines, Cosines, and Sides

In introductory physics, most of the trigonometry that you will use involves only right angles. However, in some situations, it may be easier to use one of the following rules that relate the side angles with angles for any triangle. Let a, b, and c be the sides of any triangle, and let A, B, and C be the three angles opposite these sides, respectively, as shown in Figure 5.8.

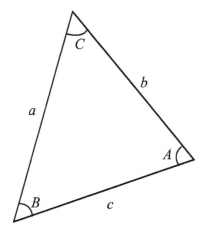

Figure 5.8

Then, according to the *law of sines*,

$$\frac{a}{\sin A} = \frac{b}{\sin B} = \frac{c}{\sin C}.$$

According to the *law of cosines*,

$$a^2 = b^2 + c^2 - 2bc \cos A,$$

$$b^2 = a^2 + c^2 - 2ac \cos B,$$

$$c^2 = a^2 + b^2 - 2ab \cos C.$$

The law of sines and the law of cosines can be proved easily using the concepts of vector cross product and vector dot product, respectively. These are left as exercies in Chapter 6.

It is not difficult to remember the law of cosines. It is the Pythagorean theorem plus the cross term involving two side lengths and the cosine of the included angle.

According to the *law of sides*,

$$a = b \cos C + c \cos B,$$

$$b = a \cos C + c \cos A,$$

$$c = a \cos B + b \cos A.$$

Example 5.7.1

Mr. Johnson walks 2.0 km due north, then makes a 120° right turn and walks for 2.5 km along a straight line. How far is Mr. Johnson from the origin of the trip? What is the direction of his original location from his present position?

Assume Mr. Johnson starts from the point P.
In this problem, $PQ = 2.0$ km,
$QR = 2.5$ km, and
angle $NQR = 120°$.
Note that the triangle PQR is not a right triangle. The sides are marked as a, b, and c and the corresponding opposite angles are marked as A, B, and C.
Determine PR ($= a$) and the angle B.

$A = 180° - 120° = 60°$

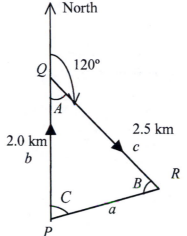

Use the law of cosines to get PR:

$$PR^2 = a^2$$

$$= b^2 + c^2 - 2bc \cos A.$$

Substitute the values. $$= (2.0 \text{ km})^2 + (2.5 \text{ km})^2 - 2(2.0 \text{ km})(2.5 \text{ km}) \cos 60°$$

$$= 4 + 6.25 - 5 = 5.25$$

Take the square root. $$a = \pm\sqrt{5.25} = \pm 2.3 \text{ km}$$

Take the positive root only. $$a = 2.3 \text{ km (since side } a \text{ must be positive)}$$

To determine the angle B, use the law of sines:

$$\frac{a}{\sin A} = \frac{b}{\sin B}.$$

Substitute the values.

$$\frac{2.3\,\text{km}}{\sin 60} = \frac{2.0\,\text{km}}{\sin B}$$

or

$$\sin B = \frac{2.0\,\text{km}}{2.3\,\text{km}} \sin 60°$$

$$= \frac{2.0\,\text{km}}{2.3\,\text{km}} (0.87) = 0.76$$

Take the *inverse sin* to get B.

$$B = \sin^{-1}(0.76) = 49.5°$$

Exercises
Use the laws of sines, cosines, or sides
to determine the following:
28. The angle A
29. The angle C
30. The side c

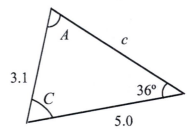

*5.8 Complex Numbers in Trigonometric Form: The Polar Form

As mentioned in Chapter 3, a complex number $z = a + bi$ is represented by a point on a two-dimensional complex plane, so that the real part of z represents the x-coordinate and the imaginary part of z represents the y-coordinate of the point in the plane. The x-axis is known as the real axis and the y-axis is known as the imaginary axis.

If (r, θ) is the location of the point in *polar coordinates* (Figure 5.9), then

$a = r \cos \theta$

$b = r \sin \theta.$

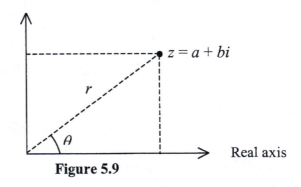

Figure 5.9

Thus, $z == a + bi = r (\cos \theta + i \sin \theta)$.

The absolute value of $z = |z| = |a + bi| = r = \sqrt{a^2 + b^2}$.

$$\text{(since } a^2 + b^2 = r^2 (\cos^2\theta + \sin^2\theta) = r^2)$$

Also, $\qquad\qquad \tan \theta = b/a$.

From the power series expansions

$$e^{i\theta} = 1 + \frac{i\theta}{1!} + \frac{(i\theta)^2}{2!} + \frac{(i\theta)^3}{3!} + \ldots$$

$$\cos \theta = 1 - \frac{\theta^2}{2!} + \frac{\theta^4}{4!} - \ldots$$

$$\sin \theta = \theta - \frac{\theta^3}{3!} + \frac{\theta^5}{5!} - \ldots$$

we obtain

$$e^{i\theta} = \cos \theta + i \sin \theta \qquad\qquad \text{and, changing } \theta \text{ to } -\theta,$$

$$e^{-i\theta} = \cos \theta - i \sin \theta.$$

From the preceding two relations we get

$$\cos \theta = \frac{e^{i\theta} + e^{-i\theta}}{2} \qquad\qquad \sin \theta = \frac{e^{i\theta} - e^{-i\theta}}{2i}.$$

Thus, from the expression $re^{i\theta} = r \cos \theta + i r \sin\theta$ it follows that a complex number $z = a + ib$ can be written in the form $z = re^{i\theta}$ if we choose r and θ so as to satisfy $a = r \cos \theta$ and $b = r \sin \theta$. The preceding expressions are useful in the mathematical analyses of many expressions that involve $\sin \theta$ and $\cos \theta$.

Exercises

Find the trigonometric form (polar form) of the complex numbers:

31. $-4i$ \qquad\qquad\qquad 32. $-\dfrac{1}{2} + \dfrac{\sqrt{3}}{2} i$

Answers to Exercises in Chapter 5

1. $h_a = 1.7, h = 3.5$

2. $h_o = 0.9, h = 1.7$

3. 20 cm

4. 7.7 m

5. 713 m

22. $\cos^{-1}(0.56) = 55.9°$, $\tan^{-1}(2.5) = 68.2°$, $\sin^{-1}(0.92) = 66.9°$

23. 37°

24. 66°

25. $\sqrt{1-x^2}$, x, $\sqrt{1-x^2}\big/x$, $x\big/\sqrt{1-x^2}$, $1\big/\sqrt{1-x^2}$, $1/x$

26. $x\big/\sqrt{1+x^2}$, $1\big/\sqrt{1+x^2}$, x, $1/x$, $\sqrt{1+x^2}\big/x$, $\sqrt{1+x^2}$

28. $A = 71°$

29. $C = 71°$

30. $c = 5.0$

31. $4(\cos 3\pi/2 + i \sin 3\pi/2) = 4e^{3\pi i/2}$

32. $1(\cos 2\pi/3 + i \sin 2\pi/3)] = 1e^{2\pi i/3}$

Chapter 6
Vectors in Physics

Vectors are introduced in physics to represent certain physical quantities and to determine the direction of a change. The physical laws can be expressed in simple forms in vector notation. The different concepts of vector algebra are described.

Introduction to Vectors

6.1   Scalars and Vectors

Physical quantities are generally of two types:

- Scalar

- Vector

A quantity, such as the volume of a container, temperature of air, time interval between two events, distance of travel, and speed of a moving object, that can be expressed by a number and its unit is called a *scalar*. However, in some cases, a direction is also needed for the complete specification of a quantity. Examples include the displacement, velocity, and acceleration of an object. These are explained in physics texts and briefly described here.

Distance and Speed as Examples of Scalars
The distance of an object from a chosen point of reference at any time is the total length of travel, and it is always positive. For a moving object, its distance changes with time. The time rate at which the distance of a moving object changes is called its *speed*. The *average speed* of a moving object is defined as

$$\text{average speed} = \frac{\text{distance traveled}}{\text{total time to travel the distance}} = \frac{d}{\Delta t} = \frac{d}{t_1 - t_2}$$

where d is the distance traveled and Δt ($= t_2 - t_1$) is the total time elapsed in traveling that distance. The SI unit for the speed is meters per second (m/s). Since *distance* is a *scalar* quantity, *speed* is also a *scalar*.

Displacement, Velocity, Acceleration, and Force as Examples of Vectors
If you walk 3 miles due north and 4 miles due east, you have traveled 7 miles but you are only 5 miles from your starting point (Figure 6.1a). That is, the magnitude of your displacement is 5 miles. The reason you cannot add in the ordinary way is that the

displacement has *direction* and *magnitude*. The *displacement* of a moving object is the straight-line distance between its final position and its initial position, along with the *direction* from the starting position to the final position.

Figure 6.1

If the initial and final positions of an object are x_1 and x_2 at time t_1 and t_2, respectively, the displacement is (Figure 6.1b)

$$\Delta x = x_2 - x_1.$$

In one dimension, there are only two possible directions of motion (left or right or up and down). The x-axis is commonly used for horizontal motions and a plus (+) sign is used to indicate the direction to the right and a minus (−) sign is used to indicate the direction to the left. Displacement is expressed in meters and it can be positive or negative. Because of its direction, displacement is a *vector* quantity.

The *velocity* of a moving object tells how fast the object is moving *and* in what direction. The average velocity is the displacement divided by the total travel time:

$$\text{average velocity} = \frac{\text{displacement}}{\text{total travel time}} = \frac{\Delta x}{\Delta t} = \frac{x_2 - x_1}{t_2 - t_1}.$$

The SI unit of velocity is meters per second (m/s). Since velocity has a direction, it is a vector quantity.

The *velocity* of a moving object changes when the object speeds up (such as when you press the accelerator of your car) or slows down (such as when you step on the brake) or when the direction of motion of the car changes (such as when you make a turn). The time rate of change of velocity of an object is called its *acceleration*. The *average acceleration* is defined as the change in velocity divided by the time taken to make the change:

$$\text{average acceleration} = \frac{\text{change in velocity}}{\text{time to make the change}} = \frac{\Delta v}{\Delta t} = \frac{v_2 - v_1}{t_2 - t_1}.$$

The SI unit of velocity is meters per second per second (m/s^2). Since velocity is a vector quantity, *acceleration* is also a *vector* quantity.

Force is simply a push or pull. It causes changes in motion or, in other words, causes an acceleration of an object. Force (**F**) is related to acceleration (**a**) by Newton's second law of motion as **F** = *m***a**, where *m* is the mass of an object. Since acceleration is a vector quantity, force is also a *vector* quantity.

Note: There are other vector quantities in physics. Examples include linear momentum, angular velocity, angular momentum, torque, electric field, and magnetic field.

A *vector* is a physical quantity that has both a magnitude and a direction. Since vectors and scalars are different types of physical quantities, they *cannot be* equal to each other. To distinguish them, different notations are used. Scalars are represented by a single symbol in *italics*. **Boldface** is used to indicate a vector with a written symbol, and *italic* is used for its magnitude. An arrow is drawn to indicate a vector on a graph. The length of the arrow is proportional to the magnitude of the vector, and the direction of the arrowhead specifies the direction of the vector. When writing vectors by hand, draw an arrow over the symbol, such as $\vec{A}$.

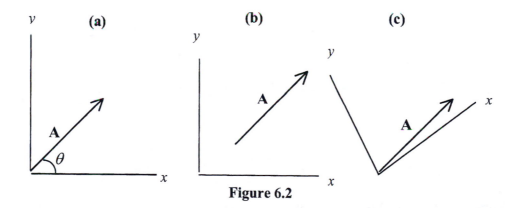

Figure 6.2

Figure 6.2 a shows a two-dimensional vector **A** whose magnitude is *A* and whose direction (as shown by the direction of the arrowhead) is specified by the angle θ made with the positive *x*-axis. The magnitude of a vector is always a positive number, but the vector can be pointed any direction, depending on the direction or the angle θ. Note that although a vector can be visualized on coordinate axes, it is independent of the coordinate system. In other words, the magnitude or direction of a vector does not change if you change the coordinate system or move the vector from one place to the other in the same coordinate system (Figure 6.2b, c). Thus, you can move a vector anywhere keeping its magnitude and direction the same.

Since a vector is characterized by both its magnitude and direction, *two vectors* are said to be *equal* when both their magnitudes *and* directions are equal.

6.2   Components of a Vector

Rectangular Vector Components

By rectangular components of a vector we mean those at right angles to each other. Suppose that the two vectors **A** and **B** are at right angles. Then the magnitude of the resultant vector **C** (Figure 6.3 a) is given by the Pythagorean theorem:

$$C = \sqrt{A^2 + B^2}.$$

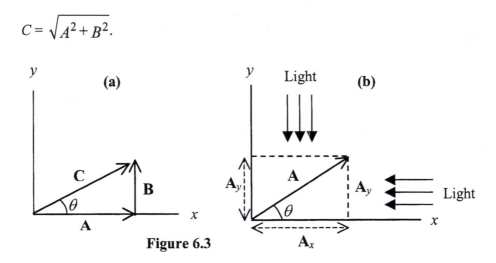

Figure 6.3

The orientation of **C** with respect to the *x*-axis is given by the inverse tangent relationship

$$\theta = \tan^{-1} \frac{B}{A}.$$

In this case, **C** is the resultant vector and **A** and **B** are called the component vectors.

Resolving a Vector into Rectangular Components

Resolving a vector into rectangular components is the reverse of adding the components. To find the components of a vector, you need to set up a coordinate system, say, a Cartesian system. If a vector **A** makes an angle θ relative to the positive *x*-axis of the coordinate system (Figure 6.3b), the *x*- and *y*-components of the vector are the projections of the vectors on the *x*- and *y*-axes, respectively. To **visualize** the vector components, imagine that a light beam is shining on the vector **A** from above the *x*-axis. Then the length of the shadow of the vector on the *x*-axis is the *x*-component A_x

of the vector. Similarly, if you imagine that a light beam is shining on the vector **A** from the right of the *y*-axis, the shadow of **A** on the *y*-axis is the *y* component A_y of the vector.

If A_x and A_y are the *x* and *y* components, respectively, of **A,** then

$$A_x = A \cos \theta$$

and

$$A_y = A \sin \theta.$$

A_x and A_y can be positive, negative, or zero depending on θ. The sign of A_x and A_y in different coordinates are shown in Table 6.1.

Table 6.1

	Quadrant I	Quadrant II	Quadrant III	Quadrant IV
A_x	*positive*	*negative*	*negative*	*positive*
A_y	*positive*	*positive*	*negative*	*negative*

The Pythagorean theorem gives the precise value of the magnitude of the vector in terms of its components:

$$A = \sqrt{A_x^2 + A_y^2} \ .$$

The direction of the vector can be obtained from the magnitude of the components and is given by

$$\theta = \tan^{-1} \frac{A_y}{A_x} \ .$$

Note: In determining an angle by means of the inverse tangent function, a calculator always displays the smallest angle. The displayed angle may or may not be the correct angle for a problem. Always utilize the quadrant grid to check your answer when you use the inverse tangent function. If you think the larger angle is the correct angle for the problem, add 180° to the smaller angle for the answer.

Example 6.2.1

Find the components of a vector of magnitude 16 m that is applied in a direction of 30° with the positive x-axis.

In this case, the vector $\mathbf{A} = 16$ m and $\theta = 30°$. The sketch of the problem follows. Use the relationships $A_x = A \cos \theta$ and $A_y = A \sin \theta$ to determine the components A_x and A_y of $\mathbf{A}$:

$$A_x = A \cos \theta = (16 \text{ m}) \cos 30°$$
$$= 16 \times 0.866 \text{ m} = 14 \text{ m}$$

$$A_y = A \sin \theta = (16 \text{ m}) \sin 30°$$
$$= 16 \times 0.5 \text{ m} = 8 \text{ m}$$

Example 6.2.2

A force ($\mathbf{F}$) is a push or pull. It is a vector and is measured in the unit of newtons (N). The x- and y-components of a force are $F_x = -4.0$ N and $F_y = 6.0$ N. Find the magnitude and direction of the force.

We have

$$F_x = -4.0 \text{ N}$$
$$F_y = 6.0 \text{ N}$$

In this problem, since the x component is negative and the y component is positive, the force vector lies in the second quadrant. Hence, the angle θ is more than 90°.

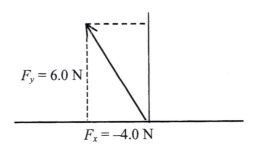

Use the Pythogorean relation and the inverse tangent relation as mentioned in Section 6.2 to solve for the vector magnitude F and the angle θ, respectively:

$$F = \sqrt{F_x^2 + F_y^2}$$

Substitute the values of F_x and F_y.

$$= \sqrt{(-4.0 \text{N})^2 + (6.0 \text{N})^2} = 7.2 \text{ N}$$

$$\theta = \tan^{-1} \frac{F_y}{F_x}$$

$$= \tan^{-1} \left(\frac{6.0 \text{ N}}{-4.0 \text{ N}} \right)$$

= −56.3° (using a calculator).

The calculator has given the smallest angle, which is in the fourth quadrant. Since the vector lies in the second quadrant, the correct angle is 180° + (−56.3°) = 123.7°.

Exercises

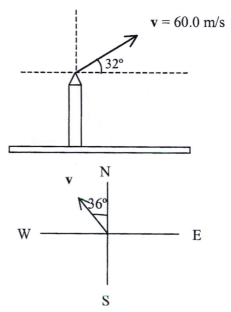

1. The velocity of a moving object is the rate of change of its displacement with time, and it is a vector. A projectile is shot from the top of a tower with a velocity of 60.0 m/s at an angle of 32° with the horizontal direction. Determine the horizontal and vertical components of the velocity vector of the projectile.

2. An airplane is traveling at a velocity $\mathbf{v}$ = 701 km/h in a direction 36° west of north. Determine the compentns of its velocity in the northernly and westerly directions.

3. The x- and y-components of a force vector are F_x = −2.0 N and F_y = −4.0 N, respectively. Determine the force vector and its direction with the positive x-axis.

4. If v_x = 2.21 units and v_y = −6.24 units, deterime the magnitude and direction of the velocity vector $\mathbf{v}$.

6.3   Unit Vectors

Unit vectors are introduced in mathematics and physics to express vectors in a convenient form. A unit vector is a dimensionless vector whose magnitude is unity and whose direction is usually one of the coordinate axes. There are *several kinds* of unit vectors in physics. Cartesian unit vectors and normal unit vectors are used in introductory physics and are described here.

Cartesian Unit Vectors

In most situations in introductory physics, you will use *Cartesian unit vectors*. Cartesian unit vectors are written as $\hat{\mathbf{i}}$, $\hat{\mathbf{j}}$, and $\hat{\mathbf{k}}$ or simply $\mathbf{i}$, $\mathbf{j}$, and $\mathbf{k}$, pointing in the positive *x*-, *y*-, and *z*-axes, respectively. The unit vectors $\mathbf{i}$, $\mathbf{j}$, and $\mathbf{k}$ are mutually perpendicular and they meet at the origin as shown in Figure 6.4.

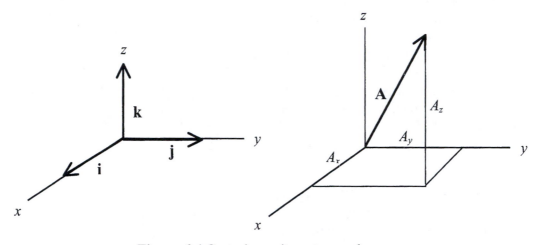

Figure 6.4 Cartesian unit vectors and Cartesian components of a vector.

An arbitrary vector **A** in space can be written as the sum of its vector components using Cartesian unit vectors as follows:

$$\mathbf{A} = A_x\mathbf{i} + A_y\mathbf{j} + A_z\mathbf{k};\ \text{in two dimensions it reduces to}\ \mathbf{A} = A_x\mathbf{i} + A_y\mathbf{j}.$$

Normal Unit Vector

The normal unit vector is usually labeled as $\hat{\mathbf{n}}$ and it is pointed outward in a direction perpendicular to the surface at the point of interest (Figure 6.5). If the surface is a plane, the normal unit vector is directed perpendicularly to the surface area. Use of the normal unit vector is common in electricity to calculate electric flux passing through an area. A small element of area *ds* has some particular orientation with respect to the coordinate axes.

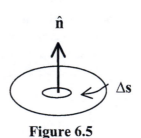

Figure 6.5

A direction has been associated with such an area along the unit vector $\hat{\mathbf{n}}$ normal to the surface, and the area is considered to be a vector ($\mathbf{ds} = ds\ \hat{\mathbf{n}}$).

Polar Unit Vectors

Polar unit vectors are labeled as $\hat{\mathbf{r}}$ and $\hat{\theta}$. The unit vector $\hat{\mathbf{r}}$ points radially outward and the unit vector $\hat{\theta}$ points perpendicularly counter-clockwise from $\hat{\mathbf{r}}$ (Figure 6.6). Unlike Cartesian unit vectors, the direction of polar unit vectors change depending on the point of interest.

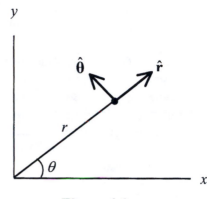

Figure 6.6

*Spherical Unit Vectors

Spherical unit vectors are labeled as $\hat{\mathbf{r}}$, $\hat{\theta}$, and $\hat{\phi}$ and they are mutually perpendicular at the point of interest. Their directions vary with the position. At any point of interest, $\hat{\mathbf{r}}$ points radially outward, $\hat{\theta}$ points perpendicularly downward from along a line tangent to the circle of radius r and $\hat{\phi}$ points perpendicularly counterclockwise from $\hat{\mathbf{r}}$ parallel to the xy-plane (Figure 6.7).

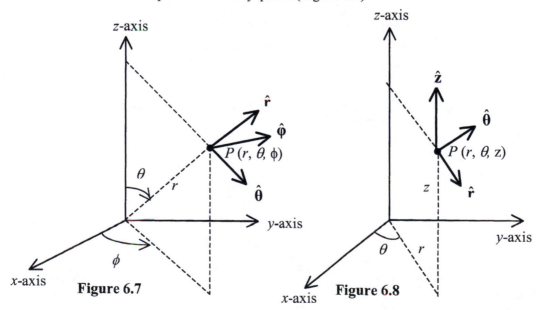

Figure 6.7

Figure 6.8

*Cylindrical Unit Vectors

Cylindrical unit vectors are labeled as $\hat{\mathbf{r}}$, $\hat{\theta}$, and $\hat{\mathbf{k}}$ and they are mutually perpendicular at the point of interest. Here $\hat{\mathbf{r}}$ and $\hat{\theta}$ are the two polar unit vectors as defined above and $\hat{\mathbf{k}}$ is the Cartesian unit vector $\hat{\mathbf{k}}$ (Figure 6.8). These unit vectors like polar and spherical polar unit vectors vary with position.

Table 6.2 lists how an arbitrary vector **A** is expressed using its components in different systems with different kinds of unit vectors.

The components of a vector in other coordinate systems are related to the Cartesian components of the vector. The relations are given in Appendix F. Also, the Cartesian unit vectors can be expressed in terms of unit vectors in other coordinate systems and vice versa. The relations between the unit vectors are given in Appendix G.

Table 6.2

Unit vector	Vector
Cartesian	$\mathbf{A} = A_x\mathbf{i} + A_y\mathbf{j} + A_z\mathbf{k}$
	$\mathbf{A} = A_x\mathbf{i} + A_y\mathbf{j}$ (in two dimensions)
Normal	$\mathbf{A} = A\,\hat{\mathbf{n}}$
Polar	$\mathbf{A} = A_r\hat{\mathbf{r}} + A_\theta\hat{\theta}$
Spherical	$\mathbf{A} = A_r\hat{\mathbf{r}} + A_\theta\hat{\theta} + A_\phi\hat{\phi}$
Cylindrical	$\mathbf{A} = A_r\hat{\mathbf{r}} + A_\theta\hat{\theta} + A_z\mathbf{k}$

Vector Addition and Subtraction

6.4   Vector Addition

Vector quantities (such as displacement, velocity, and acceleration) can be added (*vector addition*) to obtain the *resultant vector*. Either the geometric method or the analytic component method can be used to add two or more vectors.

Vector Addition: Geometric Method

Either the triangle method or the parallelogram method can be used to add two vectors geometrically.

The Triangle Method or the Tip-to-Tail Method

To add two vectors **A** and **B** by the *triangle method*, first choose a suitable scale to represent the two vectors by two arrows on graph paper, so that the magnitudes of the vectors are proportional to the lengths of the arrows and the directions of the arrowheads show the directions of the vectors. Then, place the tail of vector **B** at the tip of vector **A**. The resultant vector (the vector sum) **R** is the vector joining the tail of **A** to the head of **B**, or **R** = **A** + **B** (Figure 6.9). This method is also called the *tip-to-tail method*, since the tip of one vector is placed at the tail of the second vector.

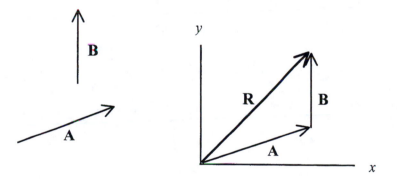

Figure 6.9

Measure the length of the resultant vector **R** (with a ruler) and the angle of the vector with the positive *x*-axis (using a protractor) for the direction. Change the length of the arrow to the magnitude of the vector by dividing by the chosen scale factor to get the magnitude of the resultant vector.

Parallelogram Method

Another graphical method, similar to the triangular method, is the *parallelogram method*. Draw the vectors **A** and **B** tail to tail, and form a parallelogram, as shown in

Figure 6.10. The diagonal represents the resultant vector **R** whose magnitude and direction can be measured from the diagram in a way similar to the triangular method.

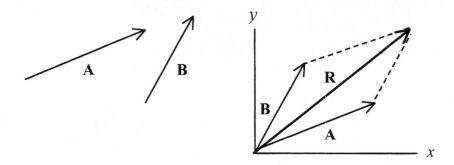

Figure 6.10

Polygon Method

The triangle method can be extended to add more than two vectors. Place all the vectors head to tail one by one, as shown in Figure 6.11, for the five vectors **A**, **B**, **C**, **D**, and **E**. The line joining the tail of the first vector **A** to the head of the last vector **E** represents the resultant vector. The method is called the polygon method, because the resulting figure is a polygon. In this case, **R** = **A** + **B** + **C** + **D** + **E**. The technique can be used to add any finite number of vectors.

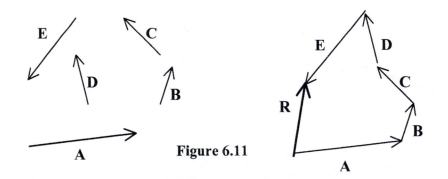

Figure 6.11

Vector Subtraction

An arrow having the same length as the original vector, but pointing in the opposite direction, represents the negative of a vector.

To subtract two vectors, **A** and **B**, add **A** to –**B**. The resultant vector **R** = **A** – **B** = **A** + (–**B**) (Figure 6.12).

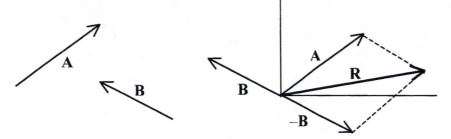

Figure 6.12

Even though the graphical methods of adding two or more vectors are useful and easy to understand, they are usually not very accurate.

Vector Addition Using Components

The *analytical component method* of vector addition involves resolving the vectors into components and then adding the components for each axis separately. If **A** and **B** are the two plane vectors to be added, and the resultant vector is **R** (= **A** + **B**), then (Figure 6.13)

$R_x = A_x + B_x$ and

$R_y = A_y + B_y.$

From the Pythagorean theorem,

$$R = \sqrt{R_x^2 + R_y^2},$$

and the angle is given by the inverse tangent relationship,

$$\theta = \tan^{-1}\left(\frac{R_y}{R_x}\right).$$

Figure 6.13

In the notation of a unit vector,
$$\mathbf{R} = \mathbf{A} + \mathbf{B} = (A_x\mathbf{i} + A_y\mathbf{j}) + (B_x\mathbf{i} + B_y\mathbf{j})$$

$$= (A_x + B_x)\mathbf{i} + (A_y + B_y)\mathbf{j}.$$

If both **A** and **B** have three components,

$$\mathbf{R} = \mathbf{A} + \mathbf{B} = (A_x\mathbf{i} + A_y\mathbf{j} + A_z\mathbf{k}) + (B_x\mathbf{i} + B_y\mathbf{j} + B_z\mathbf{k})$$

$$= (A_x + B_x)\mathbf{i} + (A_y + B_y)\mathbf{j} + (A_z + B_z)\mathbf{k}.$$

Similarly, $\mathbf{A} - \mathbf{B} = (A_x - B_x)\mathbf{I} + (A_y - B_y)\mathbf{j} + (A_z - B_z)\mathbf{k}.$

Vector addition is *commutative*: that is, $\mathbf{A} + \mathbf{B} = \mathbf{B} + \mathbf{A}$.
Vector addition is *associative*: that is, $(\mathbf{A} + \mathbf{B}) + \mathbf{C} = \mathbf{A} + (\mathbf{B} + \mathbf{C})$.

Example 6.4.1

Two vectors are given by $\mathbf{A} = 2\mathbf{i} + 3\mathbf{j} - 4\mathbf{k}$ and $\mathbf{B} = \mathbf{i} + 2\mathbf{j} + 3\mathbf{k}$. Determine the resultant vector $\mathbf{R} = \mathbf{A} + \mathbf{B}$.

We are given

$$\mathbf{A} = 2\mathbf{i} + 3\mathbf{j} - 4\mathbf{k}$$
$$\mathbf{B} = \mathbf{i} + 2\mathbf{j} + 3\mathbf{k}$$

$$\mathbf{R} = \mathbf{A} + \mathbf{B}$$

Substitute the values of **A** and **B**. $= (2\mathbf{i} + 3\mathbf{j} - 4\mathbf{k}) + (\mathbf{i} + 2\mathbf{j} + 3\mathbf{k})$

Rearrange the terms. $= (2 + 1)\mathbf{i} + (3 + 2)\mathbf{j} + (-4 + 3)\mathbf{k}$

$$= 3\mathbf{i} + 5\mathbf{j} - \mathbf{k}.$$

Example 6.4.2

A mail carrier leaves the post office and drives 20.0 mi due north to the next town. He then drives in a direction 30° north of east for 40.0 mi. Determine the displacement (**R**) of the mail carrier.

Let us choose the positive *x*-axis to be east and the positive *y*-axis north. Call the first displacement vector **A** and the second displacement vector **B**.

Since the first displacement **A** is in the *y*-direction (north), it follows that

$A_x = 0$
$A_y = 20.0$ mi.

Since the displacement **B** is 30° with the positive

x-axis (east), we have

$B_x = (40.0$ mi$) \cos 30° = (40.0$ mi$) (0.866) = 34.6$ mi.

$B_y = (40.0$ mi$) \sin 30° = (40.0$ mi$) (0.5) = 20.0$ mi.

The resultant displacement is **R**, as shown at

the right.

Add the x components to get R_x. $R_x = A_x + B_x = 0 + 34.6$ mi $= 34.6$ mi

Add the y components to get R_y. $R_y = A_y + B_y = 20.0$ mi $+ 20.0$ mi $= 40.0$ mi

Magnitude of R $R = \sqrt{R_x^2 + R_y^2} = \sqrt{(34.6 \text{ mi})^2 + (40.0 \text{ mi})^2} = 52.9$ mi

Angle θ $\theta = \tan^{-1}\left(\dfrac{R_y}{R_x}\right) = \tan^{-1}\left(\dfrac{40.0 \text{ mi}}{34.6 \text{ mi}}\right) = 49°$

You can also solve the problem using unit vector notation.

In unit vector notation, **A** $= 0\mathbf{i} + 20.0$ mi **j**   **B** $= 34.6$ mi **i** $+ 20.0$ mi **j**, and

R = **A** + **B**.

Substitute the vectors **A** and **B**. $= (0 + 34.6$ mi$) \mathbf{i} + (20.0$ mi $+ 20.0$ mi$) \mathbf{j}$

$= (34.6$ mi$) \mathbf{i} + (40.0$ mi$) \mathbf{j}$

Example 6.4.3
Relative Velocity: Raindrops are falling vertically with a velocity of 4.0 m/s. A bicyclist races through the rain with a horizontal velocity of 6.0 m/s relative to the ground. Determine the magnitude and direction of the velocity of the raindrops with respect to the bicyclist.

Since velocity is a vector quantity, the velocity of the raindrops with respect to the bicyclist is

$$\mathbf{v}_{RC} = \mathbf{v}_{RG} - \mathbf{v}_{CG},$$

where $\mathbf{v}_{RG}$ = velocity of the raindrop relative to the ground
and $\mathbf{v}_{CG}$ = velocity of the cyclist relative to the ground.

From the figure at the right, in unit vector notation,

$$\mathbf{v}_{RC} = (-6.0 \text{ m/s}) \, \mathbf{i} + (-4.0 \text{ m/s}) \, \mathbf{j}.$$

The magnitude

$$v_{RC} = \sqrt{\left(-6.0 \text{ m/s}\right)^2 + \left(-4.0 \text{ m/s}\right)^2} = 7.2 \text{ m/s}$$

The angle

$$\theta = \tan^{-1}\left(\frac{-4.0 \text{ m/s}}{-6.0 \text{ m/s}}\right) = 53°$$

Exercises

5. Let $\mathbf{A} = \mathbf{i} - 2\mathbf{j}$ and $\mathbf{B} = 2\mathbf{i} - 3\mathbf{j}$. Express $\mathbf{A} + \mathbf{B}$ and $\mathbf{A} - \mathbf{B}$ in unit vector notation. Determine the magnitude and direction of $\mathbf{A} + \mathbf{B}$ and $\mathbf{A} - \mathbf{B}$.

6. Three vectors are given by $\mathbf{A} = \mathbf{i} - 3\mathbf{j} + 4\mathbf{k}$, $\mathbf{B} = 2\mathbf{i} + 4\mathbf{j} - \mathbf{k}$ and $\mathbf{C} = 3\mathbf{i} - \mathbf{j} + \mathbf{k}$. Determine $\mathbf{A} + \mathbf{B}$, $\mathbf{A} - \mathbf{B}$, $\mathbf{B} - \mathbf{A}$, $\mathbf{A} + \mathbf{B} + \mathbf{C}$, and $\mathbf{A} - \mathbf{B} + \mathbf{C}$.

7. A jogger runs 100.0 m in a direction 30° east of north (call it displacement vector $\mathbf{A}$) and then 50.0 m in a direction 60° south of east (call it displacement vector $\mathbf{B}$). Determine the resultant displacement $\mathbf{R}$ of the jogger. Express $\mathbf{A}$, $\mathbf{B}$, and $\mathbf{R}$ in unit vector notation.

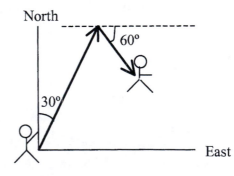

Vector Products

6.5   Products of Vectors

Multiplying a Vector by a Scalar
Multiplication of a vector by a positive scalar, s, multiplies the magnitude of the vector, but the direction of the vector does not change. If s is *negative*, the direction is *reversed* (Figure 6.14).

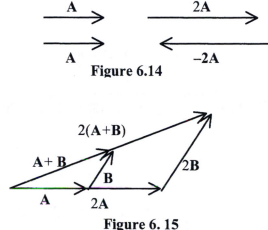

Figure 6.14

Scalar multiplication is distributive. That is, $s(\mathbf{A} + \mathbf{B}) = s\mathbf{A} + s\mathbf{B}$. This is explained in Figure 6.15 for $s = 2$.

Figure 6. 15

Note: When multiplying vectors in component form by a scalar, multiply each component of the vector by the scalar.

Scalar Product or Dot Product

The *scalar product* of two vectors is defined as the *scalar* equal to the product of the magnitudes of the vectors and the cosine of the angle between them. That is, (Figure 6.16)

$$\mathbf{A} \cdot \mathbf{B} = A\,B\cos\theta.$$

Because of the notation, the scalar product is also called the *dot product*.

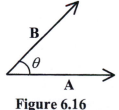

Figure 6.16

It is clear from the preceding equation that $\mathbf{A} \cdot \mathbf{B}$ can be positive, negative, or zero (zero when $\mathbf{A}$ and $\mathbf{B}$ are perpendicular, since $\cos 90° = 0$) depending on θ. Also, $\mathbf{A}^2 = \mathbf{A} \cdot \mathbf{A} = A\,A\cos 0° = A^2$. In unit vector notation,

$$\mathbf{A} = A_x\mathbf{i} + A_y\mathbf{j} + A_z\mathbf{k}, \qquad \mathbf{B} = B_x\mathbf{i} + B_y\mathbf{j} + B_z\mathbf{k}, \text{ and}$$

$$\mathbf{A} \cdot \mathbf{B} \quad = (A_x\,\mathbf{i} + A_y\,\mathbf{j} + A_z\,\mathbf{k}) \cdot (B_x\,\mathbf{i} + B_y\,\mathbf{j} + B_z\,\mathbf{k}).$$

$\mathbf{i} \cdot \mathbf{j} = \mathbf{j} \cdot \mathbf{k} = \mathbf{k} \cdot \mathbf{i} = 0$ (since the angle between each pairs is 90°)

$\mathbf{i} \cdot \mathbf{i} = \mathbf{j} \cdot \mathbf{j} = \mathbf{k} \cdot \mathbf{k} = 1$ (since the magnitudes of $\mathbf{i}$, $\mathbf{j}$, and $\mathbf{k}$ are unity)

By direct multiplication of $\mathbf{A}$ and $\mathbf{B}$, $\mathbf{A} \cdot \mathbf{B} = A_xB_x + A_yB_y + A_zB_z$.

Note: To determine the dot product, multiply like components and then add.

The dot product is *commutative*, that is, $\mathbf{A} \cdot \mathbf{B} = \mathbf{B} \cdot \mathbf{A}$.
The dot product is *distributive*, that is, $\mathbf{A} \cdot (\mathbf{B} + \mathbf{C}) = \mathbf{A} \cdot \mathbf{B} + \mathbf{A} \cdot \mathbf{C}$.

Examples of Dot Products in Physics

Work as a Dot Product

A force can cause an object to move from one place to another. Consider the displacement of a particle of mass, m, under the influence of a force as shown in Figure 6.17. In this case, a component of the force acts along the direction of the displacement, and we say that *work* is done.

Quantitatively, we define work W by the force $\mathbf{F}$ in displacing an object by $\mathbf{r}$, as the component of the force causing the displacement multiplied by the displacement. That is,

$$W = Fr \cos \theta = \mathbf{F} \cdot \mathbf{r}.$$

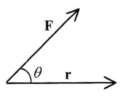

Figure 6.17

In many situations, the force varies in magnitude or direction during the displacement. For example, the force exerted by a spring increases with the amount of stretch. In general, if $\mathbf{F}$ is the force during an infinitesimal displacement $\mathbf{dr}$, the work done for this infinitesimal displacement is $\mathbf{F} \cdot \mathbf{dr}$ (Figure 6.18).

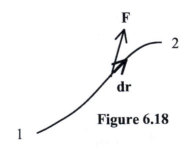

Figure 6.18

The total work in moving from point 1 and 2 is then obtained by summing all these work; that is, total work is obtained from the integration $W = \int_1^2 \mathbf{F} \cdot \mathbf{dr}$. This kind of integration is discussed in Section 9.2 of Chapter 9. A worked-out example is also given in the Section 9.2 and shows how work is calculated for a variable force.

There are other examples of dot products in introductory physics. In electrostatics, you will find that the potential energy U of an electric dipole (two equal and opposite electric charges separated by a small distance) in an electric field $\mathbf{E}$ is given by $U = -\mathbf{p} \cdot \mathbf{E}$ (where $\mathbf{p}$ is called the electric dipole moment and $\mathbf{E}$ is the electric field), and, for a magnetic dipole, the potential energy in a magnetic field $\mathbf{B}$ is $U = -\boldsymbol{\mu} \cdot \mathbf{B}$ ($\boldsymbol{\mu}$ is the magnetic dipole moment) In electricity and magnetism, the magnetic flux $d\phi$ passing through an elementary area dA is given by the dot product $d\phi = \mathbf{B} \cdot dA\,\hat{\mathbf{n}}$.

Example 6.5.1

Determine the scalar product between vector $\mathbf{A} = 2\mathbf{i} + 4\mathbf{j} + 3\mathbf{k}$ and $\mathbf{B} = 4\mathbf{i} - 2\mathbf{j} + \mathbf{k}$. Find the angle between the two vectors.

$\mathbf{A} = 2\mathbf{i} + 4\mathbf{j} + 3\mathbf{k}$ The components of $\mathbf{A}$ are $A_x = 2$, $A_y = 4$, and $A_z = 3$.

$\mathbf{B} = 4\mathbf{i} - 2\mathbf{j} + \mathbf{k}$ The components of $\mathbf{B}$ are $B_x = 4$, $B_y = -2$, and $B_z = 1$.

To determine the angle θ between the vectors, use the definition $\mathbf{A} \cdot \mathbf{B} = A\,B \cos\theta$. Knowing $\mathbf{A} \cdot \mathbf{B}$, A, and B, we can determine the angle θ.

$$\mathbf{A} \cdot \mathbf{B} = A_x B_x + A_y B_y + A_z B_z$$

Substitute the values of A_x, A_y, and A_z.
$$= (2)(4) + (4)(-2) + (3)(1) = 3$$

$$A = \sqrt{A_x^2 + A_y^2 + A_z^2}$$

Substitute the values of A_x, A_y, and A_z.
$$= \sqrt{(2)^2 + (4)^2 + (3)^2} = 5.4$$

$$B = \sqrt{B_x^2 + B_y^2 + B_z^2}$$

Substitute the values of B_x, B_y, and B_z.
$$= \sqrt{(4)^2 + (-2)^2 + (1)^2} = 4.6$$

$$\mathbf{A} \cdot \mathbf{B} = A\,B \cos\theta$$

Substitute the values of $\mathbf{A} \cdot \mathbf{B}$, A, and B. $3 = (5.4)(4.6)\cos\theta$

or

$$\cos\theta = 0.12$$

$$\theta = \cos^{-1}(0.12) = 83°$$

Example 6.5.2

Find the work done by $\mathbf{F} = 4\mathbf{i} - 2\mathbf{j} + 2\mathbf{k}$ in moving an object from a point P (3, 1, 2) to a point Q (2, –1, 6).

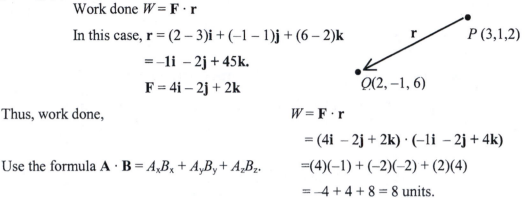

Work done $W = \mathbf{F} \cdot \mathbf{r}$

In this case, $\mathbf{r} = (2 - 3)\mathbf{i} + (-1 - 1)\mathbf{j} + (6 - 2)\mathbf{k}$

$$= -1\mathbf{i} - 2\mathbf{j} + 45\mathbf{k}.$$

$$\mathbf{F} = 4\mathbf{i} - 2\mathbf{j} + 2\mathbf{k}$$

Thus, work done, $W = \mathbf{F} \cdot \mathbf{r}$

$$= (4\mathbf{i} - 2\mathbf{j} + 2\mathbf{k}) \cdot (-1\mathbf{i} - 2\mathbf{j} + 4\mathbf{k})$$

Use the formula $\mathbf{A} \cdot \mathbf{B} = A_xB_x + A_yB_y + A_zB_z$. $= (4)(-1) + (-2)(-2) + (2)(4)$

$$= -4 + 4 + 8 = 8 \text{ units.}$$

Exercises

8. $\mathbf{A} = 2\mathbf{i} + 3\mathbf{j} - \mathbf{k}$, $\mathbf{B} = 4\mathbf{i} - 2\mathbf{j} + \mathbf{k}$, and $\mathbf{C} = \mathbf{i} + 3\mathbf{j} - 4\mathbf{k}$. Determine
    (a) the dot product $\mathbf{A} \cdot \mathbf{B}$, (b) the angle between $\mathbf{A}$ and $\mathbf{B}$, and (c) $\mathbf{A} \cdot (\mathbf{B} + \mathbf{C})$.
    Also, show that $\mathbf{A} \cdot (\mathbf{B} + \mathbf{C}) = \mathbf{A} \cdot \mathbf{B} + \mathbf{A} \cdot \mathbf{C}$.

9. Starting from the expression $\mathbf{c} = \mathbf{a} -$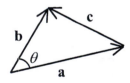
    $\mathbf{b}$, prove the law of cosines $c^2 = a^2 +$
    $b^2 - 2ab \cos\theta$, where θ is tha angle
    between the $\mathbf{a}$ and $\mathbf{b}$. (*Hint*: Find $\mathbf{c} \cdot \mathbf{c}$.)

10. Work W is defined as the dot product of the force $\mathbf{F}$ and the displacement $\mathbf{r}$. That
    is, $W = Fr \cos\theta$. Find the work done in pulling a shopping cart by a force of 20.0
    N at an angle of 30° by a distance of 2.0 m.

11. A rocket experiences a constant total thrusting force $\mathbf{F} = (200 \text{ N})\mathbf{i} + (20 \text{ N})\mathbf{j} + (120 \text{ N})\mathbf{k}$. It moves along a straight launch track and its displacement from the origin is
    $\mathbf{r} = (100 \text{ m})\mathbf{i} + (30 \text{ m})\mathbf{k}$. Determine the work done by the thursting force.

12. Two forces $\mathbf{F}_1 = (1.2\mathbf{i} - 0.5\mathbf{j} + 0.5\mathbf{k})$ N and $\mathbf{F}_2 = (0.9\mathbf{i} - 1.2\mathbf{j})$ N are applied to a
    moving object of mass 0.1 kg. The displacement produced by the forces is $\mathbf{r} = (4.0\mathbf{i} + 2.0\mathbf{j} + 2.0\mathbf{k})$ m. Determine the work done by the two forces. [*Hint*: Total
    work, $W = W_1 + W_2 = \mathbf{F}_1 \cdot \mathbf{r} + \mathbf{F}_2 \cdot \mathbf{r} = (\mathbf{F}_1 + \mathbf{F}_2) \cdot \mathbf{r}.$]

Vector Product or Cross Product

The vector product of two vectors **A** and **B** is written as **A** X **B** (Figure 6.19) and is also called a cross product. It is a vector perpendicular to both **A** and **B**, and its magnitude is defined as

$$|\mathbf{A} \times \mathbf{B}| = A\,B\sin\theta.$$

Geometrically, $|\mathbf{A} \times \mathbf{B}|$ is the area of the parallelogram generated by **A** and **B**.

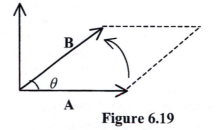

Figure 6.19

The direction of **A** X **B** (as shown in Figure 6.20) is given by the right-hand rule, which states that if the fingers of the right hand are curled in a way that will rotate **A** through the smaller angle into coincidence with **B**, then the thumb will point in the direction of **A** X **B**.

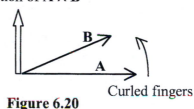

Thumb points in the direction of **A** X **B**

Curled fingers

Figure 6.20

From the definition, **A** X **A** = 0 and **A** X **B** = 0, if **A** and **B** are parallel.

Since the unit vectors **i**, **j**, and **k** are mutually perpendicular, and the cross product is perpendicular to both vectors, it follows that

i X **i** = **j** X **j** = **k** X **k** = 0 (since sin 0° = 0).

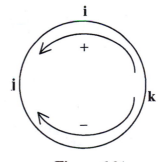

Figure 6.21

i X **j** = **k**, **j** X **k** = **i**, **k** X **i** = **j**
j X **i** = − **k**, **k** X **j** = − **i**, **i** X **k** = − **j**
Follow the direction of the arrow in Figure 6.21 to determine the sign of the products.

$$\mathbf{A} \times \mathbf{B} = (A_x\mathbf{i} + A_y\mathbf{j} + A_z\mathbf{k}) \times (B_x\mathbf{i} + B_y\mathbf{j} + B_z\mathbf{k})$$

$$= (A_yB_z - A_zB_y)\mathbf{i} + (A_zB_x - A_xB_z)\mathbf{j} + (A_xB_y - A_yB_x)\mathbf{k}$$

This can be written as an easily remembered determinant:

$$\mathbf{A} \times \mathbf{B} = \begin{vmatrix} \mathbf{i} & \mathbf{j} & \mathbf{k} \\ A_x & A_y & A_z \\ B_x & B_y & B_z \end{vmatrix}.$$

Note: To determine the cross product, form the determinant whose first row is **i, j, k**, whose second row is the components of **A**, and whose third row is the components of **B**.

To get the first term, draw a horizontal line through the first row and a vertical line through the first column. Cross multiply the remaining four terms, obtaining $A_yB_z - A_zB_y$, and then multiply by **i**.

To get the second term, draw a horizontal line through the first row and a vertical line through the second column. Cross multiply the remaining four terms, obtaining $A_xB_z - A_zB_x$, and then multiply by $-\mathbf{j}$.

To get the third term, draw a horizontal line through the first row and a vertical line through the third column. Cross multiply the remaining four terms, obtaining $A_xB_y - A_yB_x$, and then multiply by **k**.

The order of multiplication is important: $\mathbf{A} \times \mathbf{B} = -\mathbf{B} \times \mathbf{A}$ (anticommutative law of vector multiplication).

The cross product is distributive: $\mathbf{A} \times (\mathbf{B} + \mathbf{C}) = (\mathbf{A} \times \mathbf{B}) + (\mathbf{A} \times \mathbf{C})$.

Examples of Cross Products in Physics
There are several examples in physics where a physical quantity is defined as the cross product of two vectors.

Just as a force is needed to cause changes in linear motion, a *torque* is needed to cause changes in rotational motion. Torque is an example of a physical quantity that can be expressed as a cross product. *Torque* is defined as a vector given by $\tau = \mathbf{r} \times \mathbf{F}$, where $\mathbf{r}$ is the position vector of a point on an object and $\mathbf{F}$ is the force acting on the object at the point tending to rotate the object (Figure 6.22).   Since the position vector is defined with respect to a suitable point chosen as the origin, torque is defined about a point.

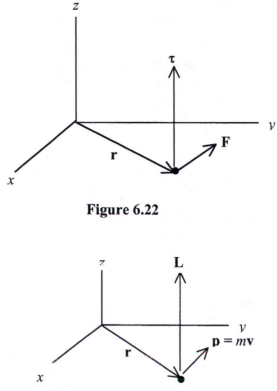

Figure 6.22

Angular momentum is another example of the cross product of two vectors. Just as an object in linear motion has a linear momentum $\mathbf{p} = m\mathbf{v}$ (where m is the mass and $\mathbf{v}$ is the velocity of the object), an object in rotational motion has a rotational momentum, called angular momentum. Angular momentum is defined as $\mathbf{L} = \mathbf{r} \times \mathbf{p}$.

If the angular velocity of a rotating object is $\boldsymbol{\omega}$ about an axis of rotation and if the position vector of a point on the object is $\mathbf{r}$ with respect to a point on the axis of rotation, then the linear velocity $\mathbf{v}$ of the point in the object is given by the cross product $\mathbf{v} = \boldsymbol{\omega} \times \mathbf{r}$.

Particles such as electrons and protons have electric charges. When an electric charge q moves with a velocity $\mathbf{v}$ in a magnetic field $\mathbf{B}$, it is found experimentally that there is a force on the moving charge. This is called magnetic force and its magnitude is found to be $F = qvB \sin \theta$. The direction of the force, as observed experimentally, is perpendicular to the plane defined by $\mathbf{v}$ and $\mathbf{B}$ (Figure 6.23). Thus, the magnitude and direction of the force on a charged particle can be expressed as the cross product

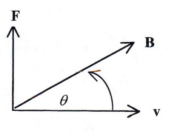

$$\mathbf{F} = q\mathbf{v} \times \mathbf{B}.$$

Figure 6.23

A current-carrying wire produces a magnetic field, which is a vector. The Biot-Savart law expresses the magnetic field $d\mathbf{B}$ at a point P distant r (Figure 6.24) from the small element $d\mathbf{l}$ of the wire along which the current I is flowing by means of the equation

Figure 6.24

$$d\mathbf{B} = \frac{\mu_0 I}{4\pi} \frac{d\mathbf{l} \times \hat{\mathbf{r}}}{r^2},$$

where $\mathbf{r}$ is the displacement vector from the element $d\mathbf{l}$ of the wire to the point P, $\hat{\mathbf{r}}$ is the unit vector along $\mathbf{r}$, and μ_0 is a constant. The magnitude of the field is $\dfrac{\mu_0 I \, dl \, \sin\theta}{4\pi r^2}$, where θ is the angle between $d\mathbf{l}$ and $\mathbf{r}$ and the direction of the field is given by the right-hand rule of the cross product. The total magnetic field at the point P due to the wire can be obtained by integrating over all current elements.

There are other examples of cross products in electricity and magnetism in introductory physics. A magnetic field exerts a force on an electric current. If the magnetic field is uniform over a straight wire of length $\mathbf{l}$, the force is $\mathbf{F} = I\mathbf{l} \times \mathbf{B}$, where I is the electric current. Also, a magnetic field tries to rotate a current loop. The torque $\boldsymbol{\tau}$ on a current loop in a magnetic field $\mathbf{B}$ is given by $\boldsymbol{\tau} = \boldsymbol{\mu} \times \mathbf{B}$, where $\boldsymbol{\mu}$ is called the magnetic dipole moment of the current loop, which is related to the number of turns, the current, and the area of the loop.

Example 6.5.1

Suppose the position vector is in the xy-plane and is given by $\mathbf{r} = 1.8 \text{ m } \mathbf{i} + 1.1 \text{ m } \mathbf{j}$. If the force vector is $\mathbf{F} = 110 \text{ N } \mathbf{j}$, determine the torque vector.

We have

$$\mathbf{r} = 1.8 \text{ m } \mathbf{i} + 1.1 \, \mathbf{j} \qquad x = 1.8 \text{ m}, y = 1.1 \text{ m}, z = 0 \text{ m}$$

$$\mathbf{F} = 110 \text{ N } \mathbf{j} \qquad F_x = 0 \text{ N}, F_y = 110 \text{ N}, F_z = 0 \text{ N}.$$

In determinant form,

$$\boldsymbol{\tau} = \mathbf{r} \times \mathbf{F} = \begin{vmatrix} \mathbf{i} & \mathbf{j} & \mathbf{k} \\ A_x & A_y & A_z \\ B_x & B_y & B_z \end{vmatrix}$$

$$= \begin{vmatrix} \mathbf{i} & \mathbf{j} & \mathbf{k} \\ x & y & z \\ F_x & F_y & F_z \end{vmatrix}.$$

Substitute the values.

$$= \begin{vmatrix} \mathbf{i} & \mathbf{j} & \mathbf{k} \\ 1.8\,\text{m} & 1.1\,\text{m} & 0 \\ 0 & 110\,\text{N} & 0 \end{vmatrix}$$

Expand the determinant.

$$= 0\mathbf{i} + 0\mathbf{j} + (198\,\text{m} \cdot \text{N})\mathbf{k} = (198\,\text{m} \cdot \text{N})\mathbf{k}$$

Example 6.5.2

The angular velocity of a rotating object about an axis of rotation is given by $\boldsymbol{\omega} = (4\mathbf{i} + \mathbf{j} -2\mathbf{k})$ rad/s. Determine the linear velocity of a point of the object having position vector with respect to a point on the axis of rotations given as $\mathbf{r} = (\mathbf{i} + 2\mathbf{j} - \mathbf{k})$ m. We have $\mathbf{v} = \boldsymbol{\omega} \times \mathbf{r}$:

$$\boldsymbol{\omega} = (4\mathbf{i} + \mathbf{j} -2\mathbf{k})\ \text{rad/s}$$
$$\mathbf{r} = (\mathbf{i} + 2\mathbf{j} - \mathbf{k})\ \text{m}.$$

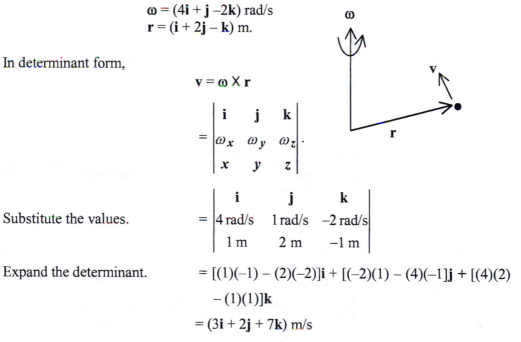

In determinant form,

$$\mathbf{v} = \boldsymbol{\omega} \times \mathbf{r}$$

$$= \begin{vmatrix} \mathbf{i} & \mathbf{j} & \mathbf{k} \\ \omega_x & \omega_y & \omega_z \\ x & y & z \end{vmatrix}.$$

Substitute the values.

$$= \begin{vmatrix} \mathbf{i} & \mathbf{j} & \mathbf{k} \\ 4\,\text{rad/s} & 1\,\text{rad/s} & -2\,\text{rad/s} \\ 1\,\text{m} & 2\,\text{m} & -1\,\text{m} \end{vmatrix}$$

Expand the determinant.

$$= [(1)(-1) - (2)(-2)]\mathbf{i} + [(-2)(1) - (4)(-1)]\mathbf{j} + [(4)(2) - (1)(1)]\mathbf{k}$$

$$= (3\mathbf{i} + 2\mathbf{j} + 7\mathbf{k})\ \text{m/s}$$

Example 6.5.3

A circular loop of wire of radius R carries a current I as shown. Use the Biot-Savart law to determine the magnetic field at the center of the wire.

Use the expression $\mathbf{dB} = \dfrac{\mu_0 I}{4\pi} \dfrac{\mathbf{dl} \times \hat{\mathbf{r}}}{r^2}$ to calculate magnetic field due to a small current element. Recall that a tangent line drawn to a circle at any point to a circle is perpendicular to radius of the circle drawn at the point.

Thus, each element $\mathbf{dl}$ of the circular loop is perpendicular to the radius vector $\mathbf{R}$, so that $\mathbf{dl} \times \hat{\mathbf{r}} = dl \sin 90^\circ = dl$.

Thus the magnetic field due to the current element $I\,\mathbf{dl}$:

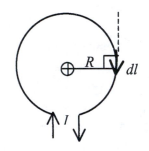

$dB = \dfrac{\mu_0 I}{4\pi} \dfrac{\mathbf{dl} \times \hat{\mathbf{r}}}{r^2} = \dfrac{\mu_0 I}{4\pi} \dfrac{dl}{R^2}$. According

to the right-hand rule, the direction of the magnetic field due to each elment $\mathbf{dl}$ of the loop is into the paper at the center.

The total magnetic field is obtained by integrating (summing) the field due to each elemtent.

$$B = \frac{\mu_0 I}{4\pi R^2} \int dl = \frac{\mu_0 I}{4\pi R^2}(2\pi R) = \frac{\mu_0 I}{2R}$$

*Triple Product of Vectors
    The cross product of two vectors $\mathbf{A}$ and $\mathbf{B}$ is a vector, and the product can be dotted or crossed with a third vector $\mathbf{C}$ to form of triple product.

Scalar Triple Product: $\mathbf{A} \cdot (\mathbf{B} \times \mathbf{C})$
    The scalar triple product is a scalar. Geometrically, it is the volume of the parallelepiped generated by the vectors $\mathbf{A}$, $\mathbf{B}$, and $\mathbf{C}$ because $|\mathbf{B} \times \mathbf{C}|$ is the area of the base and $|\mathbf{A}\cos\theta|$ is the altitude of the parallelopiped (Figure 6.25).

Figure 6.25

$\mathbf{A} \cdot (\mathbf{B} \times \mathbf{C}) = \mathbf{B} \cdot (\mathbf{C} \times \mathbf{A}) = \mathbf{C} \cdot (\mathbf{A} \times \mathbf{B})$ since they all represent the same volume of the parallopiped.

Vector Triple Product: $A \times (B \times C)$

The vector triple product is a vector and we can show that it can be written by the so called BAC-CAB rule as $A \times (B \times C) = B(A \cdot C) - C(A \cdot B)$.

*Vector Identities

Vector identities are useful in solving vector problems. Important vector identities, some of which already mentioned, are listed in Table 6.3.

Table 6.3

$s(A + B) = sA + sB$	$A \cdot (B + C) = A \cdot B + A \cdot C$
$(r + s)A = rA + sA$	$A \cdot (B \times C) = B \cdot (C \times A) = C \cdot (A \times B)$
$s(A \cdot B) = (sA) \cdot B = A \cdot (sB)$	$A \times (B + C) = (A \times B) + (A \times C)$
$s(A \times B) = (sA) \times B = A \times (sB)$	$A \times (B \times C) = B(A \cdot C) - C(A \cdot B)$

Exercises

13. $A = 2i + 3j - k$ and $B = i - 2j + 3k$, determine $A \times B$ and $B \times A$. Find the angle between A and B.

14. $A = 2i + j - 2k$, $B = 2i - j + 3k$ and $C = i + 3j - 2k$. Verify by direct calculation of right- and left-hand sides that (i) $A \times (B + C) = (A \times B) + (A \times C)$, and (ii) $A \times (B \times C) = B(A \cdot C) - C(A \cdot B)$.

15. Use vector cross product to express the area of a triangle in three different ways. Hence, prove the law of sines:

$$\frac{a}{\sin A} = \frac{b}{\sin B} = \frac{c}{\sin C}$$

16. A particle is located at $r = (2.0 \text{ m})i + (1.0 \text{ m})j + (3.0 \text{ m})k$. A force $F = (2.0 \text{ N})j - (1.0 \text{ N})k$ acts on the particle. Determine the torque about the origin.

17. A particle has a momentum $p = (4i + 6j)$ kg $\cdot$ m/s. Find its angular momentum relative to the origin when it is at $r = (2j + 5k)$ m.

18. A proton in an accelerator is propelled at 2.0×10^6 m/s perpendicular to a uniform magnetic field of 1.2 T (1 T = 1 N/(m/s)/C). Determine the force experienced by the proton. The charge for the proton, $q = 1.6 \times 10^{-19}$ C. (*Hint*: $\mathbf{F} = q\, \mathbf{v} \times \mathbf{B}$.)

19. Show that $\mathbf{A} \cdot (\mathbf{B} \times \mathbf{C}) = \begin{vmatrix} A_x & A_y & A_z \\ B_x & B_y & B_z \\ C_x & C_y & C_z \end{vmatrix}$.

20. Show that $[\mathbf{A} \times (\mathbf{B} \times \mathbf{C})] + [\mathbf{B} \times (\mathbf{C} \times \mathbf{A})] + [\mathbf{C} \times (\mathbf{A} \times \mathbf{B})] = 0$.

Answers to Exercises in Chapter 6

1. x-component = 50.9 m/s, y-component = 31.8 m/s

2. 567 km/h in the northernly direction, 412 km/h in the westernly direction

3. $F = 4.5$ N, $\theta = 243^\circ$

4. 6.6 units, $\theta = 70.5^\circ$ below the positive x-axis.

5. **A + B = 3i – 5j, A – B = – i + j**

6. **A + B = 3i + j + 3k, A – B = – i – 7j + 5k, B – A = i +7j – 5k,**
    A + B + C = 6 i + 4k, A – B + C = 2 i – 8j + 6k

7. $R = 86.6$ m, $\theta = 30^\circ$ north of east, **R = 75.0 i + 43.3 j**

8   (a) **A · B** = 1   (b) $\theta = 87^\circ$ (c) **A · (B + C)** = 16

10. 34.6 N · m

11. 2.4×10^4 N

12. 6.0 N · m

13. **A ✕ B = 7i – 7j – 7k, B ✕ A = –7i + 7j + 7k**, angle $\theta = 60^\circ$

16. **(–7.0i + 2.0j + 4.0 k)** N · m

17. **(–30i + 20j – 8k)** kg · m^2/s

18. 3.8×10^{-13} N

Chapter 7
Differential Calculus

7.1 Introduction to Calculus: A Way of Probing the Changing World

Calculus is the branch of mathematics that studies changes. Many physical quantities, which you will study in physics, change. Examples include the following: For a moving object, its position changes with time; for a rotating object, its angular displacement changes with time; for a radioactive substance, its number of radioactive nuclei changes with time; for an ideal gas kept in a closed container, its pressure changes with temperature and so on. Calculus not only is the most important tool that you will need to study such changes, many fundamental principles of physics can be conveniently expressed in simple equations using the notations of calculus.

Calculus is subdivided into two different parts: *differential* calculus and *integral* calculus. Just as the operations addition and subtraction are inverse and complementary, the two operations called differentiation in differential calculus and integration in integral calculus are inverse and complementary. Differential calculus is the study of rates of changes. More definitely, it is the study of the rate of change of one variable quantity with respect to a second variable quantity.

Although calculus can be challenging for some students, the fact is that physics would have been more difficult without the use of calculus. It is therefore important to understand the basic operations and principles of calculus before taking a calculus-based physics course. The calculus is based on two concepts of outstanding importance, namely the concept of *function* and the concept of *limit*.

7.2 Functions: Charting the Changes

Many problems in physics lead to the study of two related variables. For example, the period of a pendulum depends on the length of the pendulum, the kinetic energy of a gas depends on its velocity, and so on. When such dependence exists between two variables, either one of them is said to be a *function* of the other. Mathematically, a variable y is said to be a function of a variable x, if, for a definite value of x, a definite value of y is determined. A function of x is usually denoted by the form, $f(x)$, which is read as "the f-function of x."

For example, if $y = f(x) = x^2$, the f-function of any variable or a constant is obtained by substituting the variable or the constant for x. In this example, x is called the *independent* variable and y is called the *dependent* variable.

Thus, $f(2) = 2^2 = 4,$

$$f(z) = z^2,$$

$$f(x + b) = (x + b)^2 = x^2 + 2bx + b^2, \text{ etc.}$$

Single-Valued and Multiple-Valued Functions

A function is said to be *single valued* if it has only one value for each value of the independent variable. For example, all polynomial functions of the form $f(x) = x^n + \cdots + ax^2 + bx + c$ that contain only integral exponents are single valued. A function is said to be *multiple valued* if it has more than one value for each value of the independent variable. For example, $f(x) = \pm \sqrt{x}$ is a multiple valued function because it has two values corresponding to each value of x.

Constant Function

A function that has the same value $f(x) = a$ for all values of x in an interval is called a *constant* function. A constant function is graphically represented by a straight line parallel to x-axis as shown in Figure 7.1.

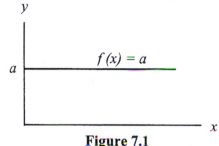

Figure 7.1

Monotonic Increasing and Monotonic Decreasing Functions

A function $f(x)$ for which an increase in the value of x throughout the interval causes an increase in the value of $f(x)$, is called a *monotonic increasing* function (Figure 7.2). On the other hand, a function $f(x)$ for which an increase in the value of x always causes a decrease in the value of $f(x)$ is called a *monotonic decreasing* function (Figure 7.3).

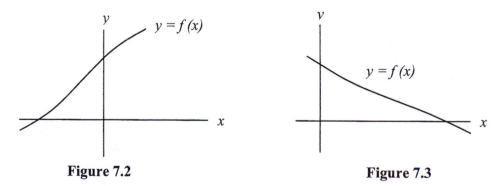

Figure 7.2   **Figure 7.3**

Even and Odd Functions

A function $f(x)$ is said to be *even* if $f(-x) = f(x)$. The curve represented by an even function $y = f(x)$ is symmetrical with respect to the y-axis. From the plot of $y = \cos x$

(Figure 7.4a.) it is seen that cos x is an even function of x. The function $y = f(x) = x^2$ is also an even function of x, as shown in Figure 7.4b.

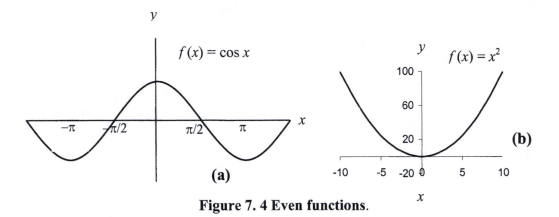

Figure 7. 4 Even functions.

A function $f(x)$ is said to be *odd* if $f(-x) = -f(x)$. The curve represented by an odd function is symmetrical with respect to the origin. From the plot of $y = \sin x$ (Figure 7.5a) it is seen that sin x is an odd function of x. The functions $y = x$ and $y = x^3$ (Figure 7.5b) are also odd. The concept of even and odd functions is important in introductory quantum physics.

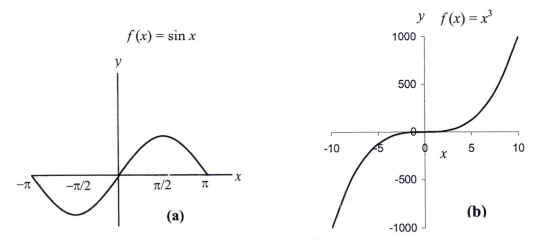

Figure 7. 5 Odd functions.

7.3   Limit and Continuity

Limit

The whole structure of calculus is based on a concept called the *limit*. By the statement "the variable x approaches the limit a," or "x tends to a" in calculus, we mean that x assumes successive values whose numerical differences from a become gradually less and less and finally becomes less than any small quantity that we can imagine. This is denoted by the symbol $x \rightarrow a$. The successive values of x can be greater than or less than a.

As x approaches the number a (*but is not equal to a*), if there exists a finite number L toward which the function $f(x)$ approaches such that the numerical difference between $f(x)$ and L becomes as small as we can imagine by approaching x very close to a, then L is called the limit of the function $f(x)$ as x tends to a. This is mathematically written as $\lim\limits_{x \to a} f(x) = L$.

Important Note: Consider the function $f(x) = \dfrac{x^2 - 4}{x - 2}$. The function $f(x)$ is undefined at $x = 2$; that is, $f(2)$ does not exist since the denominator becomes zero for $x = 2$. But $\lim\limits_{x \to 2} f(x) = \lim\limits_{x \to 2} \dfrac{x^2 - 4}{x - 2} = \lim\limits_{x \to 2} \dfrac{(x+2)(x-2)}{x-2}$

$= \lim\limits_{x \to 2} (x + 2) = 2 + 2 = 4$. Note that in this case we could cancel $x - 2$ from both the numerator and the denominator since $x - 2$ is not zero because $x \neq 2$.

Continuity and Discontinuity of Functions

A function $f(x)$ is said to be *continuous* if the plot of $f(x)$ is a continuous curve. But if there is a break or discontinuity at a point on the curve of $f(x)$, the function $f(x)$ is discontinuous for the corresponding value of x. For example, consider two functions of x: $f(x) = x$ and $g(x) = x^2/x$. Their plots are shown in Figures 7.6 and 7.7, respectively. Note that $f(x)$ is a continuous function of x passing through $x = 0$. But the function $g(x)$ is discontinuous at $x = 0$ because for $x = 0$, $g(x) = x^2/x$ is undefined, as its denominator is zero for $x = 0$.

Mathematically, a function $f(x)$ is said to be continuous at a point $x = a$ if $\lim\limits_{x \to a} f(x)$ exists, is finite, and is equal to the value $f(a)$.

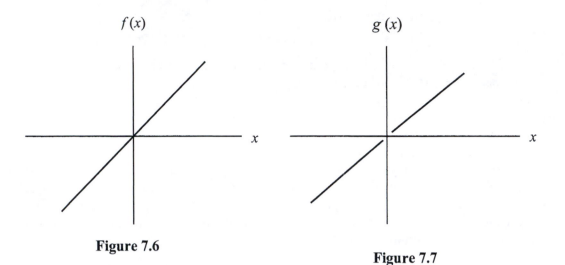

Figure 7.6

Figure 7.7

7.4   Derivative as the Rate of a Change

Let y be a function of the variable x, that is, $y = f(x)$. Let Δy be the change in y corresponding to a change Δx of x. This can be written as $y + \Delta y = f(x + \Delta x)$. Then the derivative of y or $f(x)$ with respect to x is mathematically defined as the limit

$$\frac{dy}{dx} = \lim_{\Delta x \to 0} \frac{\Delta y}{\Delta x} = \lim_{\Delta x \to 0} \frac{f(x + \Delta x) - f(x)}{\Delta x}.$$

For example, consider the function $y = f(x) = x^2$. From the definition,

$$\frac{dy}{dx} = \lim_{\Delta x \to 0} \frac{(x + \Delta x)^2 - x^2}{\Delta x} = \lim_{\Delta x \to 0} \frac{x^2 + 2x\,\Delta x + (\Delta x)^2 - x^2}{\Delta x}$$

$$= \lim_{\Delta x \to 0} \frac{\Delta x(2x + \Delta x)}{\Delta x} = \lim_{\Delta x \to 0} (2x + \Delta x) = 2x.$$

The derivative of the function $y = f(x)$ is also written as y', df/dx, or $f'(x)$. The process of finding a derivative is called *differentiation*. An important aspect of derivative that follows from the preceding definition is that the *derivative is the rate of change of one variable with respect to the other; that is, dy/dx is the rate of change of y with respect to x*. This will be clear from the following examples of physics.

Speed, Velocity, and Acceleration as Derivatives (Time Rate of Change)

For an object moving along the x-axis, its position x changes with time t. That is, the position x is a function of time t. If x and $x + \Delta x$ represent the position of the object at time t and $t + \Delta t$, respectively, then Δx denotes the distance covered by the object in time Δt (Figure 7.8).

Figure 7.8

The ratio $\Delta x / \Delta t$ in the limit Δt tends to zero represents the instantaneous rate at which the object is describing its path in unit time. On the other hand, by definition, dx/dt is the derivative of x with respect to t. Thus, the derivative dx/dt represents the instantaneous rate of change of x with respect to t, or the *instantaneous speed*, v, of the object. That is,

$$v = \lim_{\Delta t \to 0} \frac{\Delta x}{\Delta t} = \frac{dx}{dt}.$$

Instantaneous velocity of an object is its instantaneous speed with direction.

Also, if v represents the velocity of a moving object at time t, dv/dt represents the instantaneous time-rate of change of velocity, that is, its *instantaneous acceleration*, a, or,

$$a = \lim_{\Delta x \to 0} \frac{\Delta v}{\Delta t} = \frac{dv}{dt}.$$

Similarly, if the pressure of a gas is p and its temperature is T, then for the gas kept in a closed container (that is, volume constant), the derivative dp/dT represents the rate of change of pressure with temperature and so on.

Note: *In general, if y is a function of x, the derivative dy/dx represents the rate of change of y with respect to x.*

- Note that for the function $y = f(x)$, if *y increases when x is increased* and *decreases when x is decreased*, the corresponding changes Δy and Δx are of the same sign and, hence, the ratio $\Delta y / \Delta x$ and the derivative dy/dx is *positive*.

- But if *y decreases when x is increased, or increases when x is decreased*, dy/dx is *negative*.

- If y remains constant as x increases or decreases, dy/dx is zero.

Geometrical Significance of Derivatives

The geometrical significance of derivatives is represented graphically as shown in Figure 7.9. Let A be a point (x, y) on the curve and B a neighboring point $(x + \Delta x, y + \Delta y)$ that may be on either side of A so that Δx can be positive or negative. If ϕ be the inclination of the line AB with x-axis, the slope of the line AB is given by

$$\tan \phi = \frac{(y + \Delta y) - y}{(x + \Delta x) - x} = \frac{\Delta y}{\Delta x}.$$

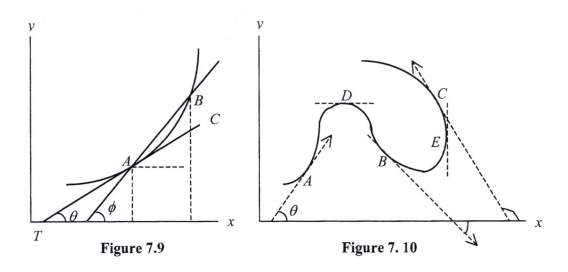

Figure 7.9   **Figure 7. 10**

Now let B approach A along the curve indefinitely closely so that $\Delta x \rightarrow 0$. As B approaches A, the line AB tends to a definite limiting position TAC, which is the tangent to the curve drawn at the point A. In this case, if θ is the inclination of the straight line TAC to the x-axis, then as AB tends to TAC, $\phi \rightarrow \theta$. Also, as $\Delta x \rightarrow 0$,

$\dfrac{dy}{dx} = \lim\limits_{\Delta x \rightarrow 0} \dfrac{\Delta y}{\Delta x}$ from the definition of the derivative. Thus,

$$\tan \theta = \frac{dy}{dx} = \text{slope of the tangent to the curve drawn at } A.$$

Therefore, *the derivative dy/dx for any value of x (when it exists) is the trigonometrical tangent of the inclination of the tangent line of the curve with positive x-axis at the point x, or the slope of the curve at the point x.*

- If dy/dx is positive (such as point A of Figure 7.10), $\tan\theta$ is positive, the angle θ is acute, and at that point y increases with x.

- If dy/dx is negative (such as point B of Figure 7.10), $\tan\theta$ is negative, the angle θ is negative or the angle θ is obtuse (such as point C Figure 7.10), and y decreases when x increases or vice versa.

- If $dy/dx = 0$ (such as point D of Figure 7.10), the tangent line is parallel to the x-axis.

- If $dy/dx \rightarrow \infty$, that is, $dx/dy = 0$, (such as point E, Figure 7.10), the tangent line is parallel to the y-axis.

7.5   Derivatives of Common Functions

To determine the derivatives of single-term common mathematical functions with respect to the variables that are in the functions, you do not need to determine the slopes or the tangent lines. These can be done from the mathematical definition of the derivative, that is, by evaluating the limit that occurs in the definition. These are done in any standard text on differential calculus. The results of these derivatives are useful in physics. Derivatives of most common single-term functions are listed below. Here c and n are constants.

$$\frac{d}{dx}(c) = 0 \qquad\qquad \frac{d}{dx}[c\,f(x)] = c\,\frac{d}{dx}[f(x)]$$

$$\frac{d}{dx}(x^n) = nx^{n-1} \qquad\qquad \frac{d}{dx}(e^{cx}) = ce^{cx}$$

$$\frac{d}{dx}(\ln cx) = \frac{1}{x} \qquad\qquad \frac{d}{dx}(\sin cx) = c\cos cx$$

$$\frac{d}{dx}(\cos cx) = -c\sin cx \qquad\qquad \frac{d}{dx}(\tan cx) = c\sec^2 cx$$

$$\frac{d}{dx}(\cot cx) = -c\csc^2 cx \qquad\qquad \frac{d}{dx}(\csc cx) = -c\csc cx \cot cx$$

$$\frac{d}{dx}(\sec cx) = c\sec cx \tan cx \qquad\qquad \frac{d}{dx}(\sin^{-1} cx) = c\,\frac{1}{\sqrt{1-(cx)^2}}$$

$$\frac{d}{dx}(\cos^{-1} cx) = c \frac{-1}{\sqrt{1-(cx)^2}}$$

$$\frac{d}{dx}(\tan^{-1} cx) = c \frac{1}{1+(cx)^2}$$

$$\frac{d}{dx}(\cot^{-1} cx) = c \frac{-1}{1+(cx)^2}$$

$$\frac{d}{dx}(\sinh cx) = c \cosh cx$$

$$\frac{d}{dx}(\cosh cx) = c \sinh cx$$

$$\frac{d}{dx}(\tanh cx) = c \, \text{sech}^2 cx$$

$$\frac{d}{dx}(\coth cx) = -c \, \text{csch}^2 cx$$

$$\frac{d}{dx}(\sinh^{-1} cx) = c \frac{1}{\sqrt{(cx)^2+1}}$$

$$\frac{d}{dx}(\tanh^{-1} cx) = c \frac{1}{1-(cx)^2}$$

$$\frac{d}{dx}(\cosh^{-1} cx) = c \frac{\pm 1}{\sqrt{(cx)^2-1}}$$

$$[+ \text{ if } \cosh^{-1} cx > 0, - \text{ if } \cosh^{-1} cx < 0]$$

In case it is not possible to remember all of the preceding derivatives, *remember* the following simple results that are often useful in physics:

The derivative of a constant = 0.

The derivative of the product of a constant and a function = product of the constant and the derivative of the function.

The derivative of $x^n = nx^{n-1}$

The derivative of $e^x = e^x$.

The derivative of $\ln x = 1/x$.

The derivative of $\sin x = \cos x$.

The derivative of $\cos x = -\sin x$.

The derivative of $\tan x = \sec^2 x$.

7.6   Rules of Derivatives

In many situations in physics you will find derivatives of functions, which are more complicated than the functions listed in the preceding section. The following rules will be useful in those situations. Here u and v are functions of x.

1. **The Addition-Subtraction Rule**: *The derivative of the sum or difference of two functions is equal to the sum or difference of their derivatives*:

$$\frac{d}{dx}[u \pm v \pm \cdots] = \frac{du}{dx} \pm \frac{dv}{dx} \pm \cdots .$$

2. **The Product Rule**: *The derivative of the product of two functions = first function × derivative of the second + second function × derivative of the first*:

$$\frac{d}{dx}[u \cdot v] = u\frac{dv}{dx} + v\frac{du}{dx} .$$

3. **The Quotient Rule**: *The derivative of the quotient of two functions =*

$$\frac{derivative \; of \; the \; num. \; × \; denom. - derivative \; of \; denom. \; × \; num}{square \; of \; denom.} .$$

$$\frac{d}{dx}\left(\frac{u}{v}\right) = \frac{v(du/dx) - u(dv/dx)}{v^2} .$$

4. **Generalized Product Rule**: *The derivative of u^n with respect to x is given by the derivative of the function times the derivative of u with respect to x*:

$$\frac{d}{dx}(u^n) = n\,u^{n-1}\frac{du}{dx} .$$

5. **The Chain Rule**: If y is a function of the variable t, $y = f(t)$, and t is a function of x, $t = f(x)$, then the derivative of y with respect to x is given by the chain rule:

$$\frac{dy}{dx} = \frac{dy}{dt} \cdot \frac{dt}{dx} .$$

Also, note that $\dfrac{dt}{dx} = \dfrac{1}{dx/dt}$ (provided neither derivative is zero).

Thus, $\dfrac{dy}{dx} = \dfrac{dy/dt}{dx/dt}$.

The preceding expression for dy/dx is useful to calculate derivative of parametric equations where both x and y can be expressed separately in terms of the parameter t.

Example 7.6.1

Find the derivative of the function $f(x) = 2x^5 - 4x^3 + 2x^2 + 6$ with respect to x.

$$f'(x) = \frac{d}{dx}(2x^5 - 4x^3 + 2x^2 + 6)$$

Use the addition-subtraction rule.
$$= \frac{d}{dx}(2x^5) - \frac{d}{dx}(4x^3) + \frac{d}{dx}(2x^2) + \frac{d}{dx}(6)$$

Take the constants out.
$$= 2\frac{d}{dx}(x^5) - 4\frac{d}{dx}(x^3) + 2\frac{d}{dx}(x^2) + \frac{d}{dx}(6)$$

Use the rules for derivative, $\dfrac{d}{dx}(c) = 0$ and $\dfrac{d}{dx}(x^n) = nx^{n-1}$.

Thus,
$$f'(x) = 2(5x^{5-1}) - 4(3x^{3-1}) + 2(2x^{2-1}) + 0$$
$$= 10x^4 - 12x^2 + 4x$$

Factor out $2x$.
$$= 2x(5x^3 - 6x + 2).$$

Example 7.6.2

A prticle moves along the x-axis and is postion x changes with time t according to the equation $x = (4.0 \text{ m})\, t + (2.4 \text{ m/s}^2)\, t^2$. Determine the velocity $v = dx/dt$ and acceleration $a = dv/dt$ of the particle as a function of time.

$$x = (4.0 \text{ m/s})\, t + (2.4 \text{ m/s}^2)t^2$$

Take the derivative with respect to t. $v = \dfrac{dx}{dt} = \dfrac{d}{dt}[(4.0 \text{ m/s})t + (2.4 \text{ m/s}^2)t^2]$

Use the addition-subtraction rule.
$$= \frac{d}{dt}[(4.0 \text{ m/s})t] + \frac{d}{dt}[(2.4 \text{ m/s}^2)t^2]$$

Take the constants out and use the formula $\dfrac{d}{dx}(x^n) = nx^{n-1}$.

Thus, $v = (4.0 \text{ m/s}) (1) + (2.4 \text{ m/s}^2) (2t^{2-1}) = 4.0 \text{ m/s} + (4.8 \text{ m/s}^2)t$.

Take the derivative with respect to t.

$$a = \frac{dv}{dt} = \frac{d}{dt}[(4.0 \text{ m/s})t + (4.8 \text{ m/s}^2)t]$$

$$= \frac{d}{dt}(4.0 \text{ m/s}) + \frac{d}{dt}[(4.8 \text{ m/s}^2)t]$$

$$= 0 + (4.8 \text{ m/s}^2) (1) = 4.8 \text{ m/s}^2.$$

Example 7.6.3

Find the derivative of the function $f(x) = e^x \cos x$ with respect to x.

$$f'(x) = \frac{d}{dx}(e^x \cos x)$$

Use the product rule.

$$= e^x \frac{d}{dx}(\cos x) + \cos x \frac{d}{dx}(e^x)$$

Use the results for derivatives of $\sin x$ and $\cos x$.
Factor out e^x.

$$= e^x (-\sin x) + \cos x (e^x)$$
$$= e^x (\cos x - \sin x)$$

Example 7.6.4

Find the derivative of the function $y = \dfrac{\sin x}{x^3}$ with respect to x.

$$\frac{dy}{dx} = \frac{d}{dx}\left(\frac{\sin x}{x^3}\right)$$

Use the quotient rule.

$$= \frac{x^3 \dfrac{d}{dx}(\sin x) - \sin x \dfrac{d}{dx}(x^3)}{\left(x^3\right)^2}$$

Use the results for derivatives.

$$= \frac{x^3 (\cos x) - \sin x (3x^{3-1})}{x^6}$$

$$= \frac{x^3 \cos x - 3x^2 \sin x}{x^6}$$

Factor out x^2.

$$= \frac{x^2 (x \cos x - 3 \sin x)}{x^6}$$

$$= \frac{(x\cos x - 3\sin x)}{x^4}$$

Example 7.6.5

Find $\dfrac{dy}{dx}$ if $x = a\,(t - \sin t)$, $y = a\,(1 + \cos t)$.

In this case, x and y have been expressed in terms of a third parameter t. Thus,

$$\frac{dy}{dx} = \frac{dy/dt}{dx/dt}$$

$$\frac{dx}{dt} = \frac{d}{dt}[a\,(t - \sin t)] = a\frac{d}{dt}\,(t - \sin t) = a\,(1 - \cos t)$$

$$\frac{dy}{dt} = \frac{d}{dt}[a\,(1 + \cos t)] = a\frac{d}{dt}\,(1 + \cos t) = a\,(0 - \sin t) = -\,a\sin t.$$

Therefore,

$$\frac{dy}{dx} = \frac{dy/dt}{dx/dt} = \frac{-a\sin t}{a(1 - \cos t)} = -\frac{2\sin\dfrac{t}{2}\cos\dfrac{t}{2}}{2\sin^2\dfrac{t}{2}} = -\cot\frac{t}{2}.$$

7.7   Derivatives of Implicit Functions

In many situations in physics, it may be inconvenient or even impossible to solve an equation of the form, say, $f\,(x, y) = 0$ for y in terms of x. In such cases, y is said to be an *implicit function* of x. To find derivative of such implicit functions, *take the derivative of each term* of the equation with respect to x regarding y as an unknown function of x having a derivative dy/dx. Then use algebra to solve the resulting equation for dy/dx.

Example 7.7.1

Find $\dfrac{dy}{dx}$, if $x^3 - xy^3 - 2y + 6 = 0$.

This equation cannot be solved easily for y in terms of x. Therefore, we must use implicit derivation to find the derivative. Differentiating each term of the equation with respect to x,

$$3x^2 - \left(x \cdot 3y^2 \frac{dy}{dx} + y^3 \cdot 1\right) - 2\frac{dy}{dx} + 0 = 0$$

Collect terms involving $\frac{dy}{dx}$. $(3xy^2 + 2)\frac{dy}{dx} = 3x^2 - y^3$

Divide by $(3xy^2 + 2)$. $\frac{dy}{dx} = \frac{3x^2 - y^3}{3xy^2 + 2}$

Note that the final answer contains y.

Example 7.7.2 (thermal physics)

In an isothermal process of an ideal gas PV = constant, while for an adiabatic process for an ideal gas PV^γ = constant, where γ is a constant (greater than one) of the gas and P and V represent the pressure and the volume of the gas, respectively. If V is plotted along the x-axis and P is plotted along the y-axis, compare the slope the isothermal and the adabatic curves.

For an isothermal process PV = constant.
Differentiate both sides with respect to V.

Use the product rule. $P \cdot 1 + V\frac{dP}{dV} = 0$

Solve for $\frac{dP}{dV}$. $\frac{dP}{dV} = -\frac{P}{V}$

= slope of the isothermal curve.

For an adiabatic process PV^γ = constant.

Differentiate both sides with respect to V.

Use the product rule. $P \cdot (\gamma V^{\gamma - 1}) + V^\gamma \frac{dP}{dV} = 0$

Solve for $\frac{dP}{dV}$. $\frac{dP}{dV} = -\frac{\gamma P V^{\gamma - 1}}{V^\gamma} = -\frac{\gamma P}{V}$

= Slope of the adiabatic curve.

Thus, the ratio of the two slopes = $-\frac{\gamma P}{V}\bigg/-\frac{P}{V} = \gamma\ (>1)$.

Exercises

Find the derivative of the following functions of x.

1. $3x^7 + 4x^4 - 2x^2 - x + 6$ 2. $x^2 \log x$ 3. $(x^2 + 1) \sin x$

4. $\dfrac{\sin x}{x}$ 5. $\dfrac{1 + \sin x}{1 - \sin x}$

Find $\dfrac{dy}{dx}$ for the following cases:

6. $x = a \cos t, y = b \sin t$ 7. $x = a(2\cos t + \cos 2t), y = a(2\sin t - \sin 2t)$

8. $3x^4 - x^2 y + 2y^3 = 0$ 9. $x^{2/3} + y^{2/3} = a^{2/3}$

10. Determine the velocity (v) and acceleration (a) of a particle whose position (x) changes with time according to the equation $x = (1.2 \text{ m}) + (2.8 \text{ m/s}) \, t + (4.2 \text{ m/s}^2)t^2$.

11. The force of gravity on a person of mass m at a distance r from the center of the earth is given by $F = GM_E m/r^2$, where G is a constant (called gravitational constant) and M_E is the constant mass of the earth. The force of gravity acting on a person is called his weight. If the person travels at a constant velocity v directly away from the earth, show that the time (t) rate of change of the weight of the person is $-(2GM_E m/r^3)v$. (*Hint*: $v = dr/dt$.)

7.8   Differentials

Let $f'(x)$ be the derivative of the function $y = f(x)$ and Δx be the increment or the change of x. Then the differential of y or $f(x)$, which is denoted by the symbol dy or $df(x)$, is given by

$$dy = df(x) = f'(x)\, \Delta x.$$

If $f(x) = x$, or, $f'(x) = 1$, then from the preceding equation, $dx = 1 \cdot \Delta x = \Delta x$. Therefore, if x is the independent variable, the differential of x, written as dx, is identical with Δx.

Thus, $dy = df(x) = f'(x)\, dx.$

That is, the *differential of a function is equal to the derivative of the function multiplied by the differential of the independent variable.*

Thus, if $y = x^n$, $dy = nx^{n-1}\, dx.$ If $y = \tan x,$ $dy = \sec^2 x \, dx.$

Rules for Differentials

The rules for differentials are exactly analogous to those for derivatives:

$$d(c) = 0$$

$$d(u \pm v \pm w \pm \cdots) = du \pm dv \pm dw \pm \cdots$$

$$d(uv) = u\,dv + v\,du$$

$$d\left(\frac{u}{v}\right) = \frac{v\,du - u\,dv}{v^2}$$

$$d(\sin x) = \cos x\,dx$$

$$d(\cos x) = -\sin x\,dx.$$

7.9   Higher Derivatives of a Function

It is possible to take the derivative of derivatives of a function. The second derivative of the function $y = f(x)$ is the derivative of its first derivative. It is written as follows:

$$\text{Second derivative of } y \text{ or } f(x) = \frac{d}{dx}\left(\frac{dy}{dx}\right) = f''(x) = y''.$$

For example, the velocity $v = \dfrac{dx}{dt}$ and the acceleration $a = \dfrac{dv}{dt}$. Thus, the acceleration a can be written as the second derivative of the position x with respect to time t as follows:

$$a = \frac{dv}{dt} = \frac{d}{dt}\left(\frac{dx}{dt}\right) = \frac{d^2x}{dt^2}.$$

Recall, from Section 7.4, that first derivative of a function at a point is the value of the slope of the curve representing the function at the point. Since the second derivative of a function is the derivative of its first derivative, it represents the rate at which the slope (or first derivative) changes with respect to x, or, in other words, it is a measure of how much curvature or concavity a function has. For a function that is *concaving up*, the second derivative is *positive* (Figure 7.11), while for a function that is *concaving down* (Figure 7.12), the second derivative is *negative*.

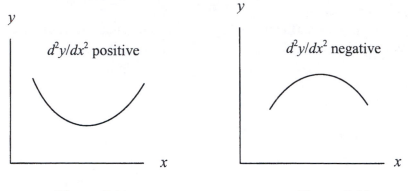

Figure 7.11   **Figure 7.12**

Higher derivatives beyond second derivatives can also be found by continuing taking derivatives of derivatives.

Third derivative of $y = f(x)$ with respect to $x = \dfrac{d}{dx}\left(\dfrac{d^2 y}{d x^2}\right)$

$$= \dfrac{d^3 y}{d x^3} = f'''(x) = y'''.$$

In general,

nth derivative of y with respect to $x = \dfrac{d}{dx}\left(\dfrac{d^{n-1} y}{d x^{n-1}}\right) = \dfrac{d^n y}{d x^n} = f^{(n)}(x) = y^{(n)}.$

Higher derivatives beyond the second derivative are rarely used in introductory or intermediate physics.

7.10   Maxima and Minima of a Function

The concept of derivative is directly used in physics to determine the maxima and minima of a function representing a physical quantity.

By the maximum value of a function $y = f(x)$ in calculus, we do not necessarily mean the absolutely greatest value attainable by the function. A function may have several maxima and minima. A function $y = f(x)$ is said to be a minimum at a point $x = a$ if $f(a)$ is less than every other value assumed by $f(x)$ in the immediate neighborhood of $x = a$. Similarly, a function $y = f(x)$ is said to be a *maximum* at a point $x = b$ if $f(b)$ is greater than every other value assumed by $f(x)$ in the *immediate neighborhood* of $x = b$. These maxima and minima points on the curve are called *critical points*.

For all critical points on the curve, the tangent line is horizontal, that is, the first derivative dy/dx or $f'(x)$ is zero. In Figure 7.13 the function $f(x)$ has a minimum at $x = a$ and $x = c$ and a maximum at $x = b$. Thus, $f'(a) = f'(b) = f'(c) = 0$. This is a *necessary condition* for maximum and minimum. You can use this condition to determine the locations of maxima and minima of a function. This technique is useful in different areas of physics especially for potential energy analyses. The maxima and minima are collectively called *extrema* on the function.

Figure 7.13

Follow these steps to determine the extrema of a function $y = f(x)$:

- Determine the first derivative dy/dx, that is, $f'(x)$.

- Set $f'(x) = 0$, and solve the resulting equation for x. The values of x are the critical points.

- Determine the second derivative $f''(x)$ and find the value of the second derivative for each critical point.

- If the result is negative, then $f(x)$ has a maximum at the corresponding critical point. Plug the value of x for which $f(x)$ has a maximum to the original function to determine the maximum value of the function $f(x)$.

- If the result is positive, then $f(x)$ has a minimum at the corresponding critical point. Plug the value of x for which $f(x)$ has a maximum to the original function to determine the maximum value of the function $f(x)$.

- If the result is zero, pick a number greater than and a number less than the critical point. The number must be close to the test point so that the value of another critical point is not inadvertently tested. Substitute these two numbers into the second derivative.

    If both numbers are negative, $f(x)$ has a maximum at the corresponding critical point.

If both numbers are positive, $f(x)$ has a minimum at the corresponding critical point.

If one number is positive and the other negative, $f(x)$ has an inflection point. At inflection points the curve change concavity. Ignore any inflection point. If an extrema is found, plug the value of x to the original function to determine the maximum or minimum value of the function $f(x)$.

Note: A point at which the slope of a curve has a maximum or minimum value is called a point of inflection. A curve bends upward if second derivative is *positive*, and bends downward if second derivative is *negative*.

Example 7.10.1
Investigate for what values of x the function $f(x) = 2x^3 - 21x^2 + 36x - 21$ is a minimum or maximum. Find also the minimum and maximum values of $f(x)$.

$$f(x) = 2x^3 - 21x^2 + 36x - 21$$

A necessary condition for maximum and minimum is that the first derivative $f'(x) = 0$. Find the first derivative $f'(x)$.

$$f'(x) = 2(3x^2) - 21(2x) + 36 + 0 = 6x^2 - 42x + 36$$

Set $f'(x) = 0$, to determine maximum or minimum. Thus,

$$6x^2 - 42x + 36 = 0$$

Factor out 6. $$6(x^2 - 7x + 6) = 0$$

Divide by 6. $$x^2 - 7x + 6 = 0$$

Factor the left-hand side. $$(x - 1)(x - 6) = 0$$

or $$x = 1 \text{ or } 6.$$

Determine the second derivative $f''(x)$.

$$f''(x) = 6(2x) - 42 = 12x - 42 = 6(2x - 7)$$

For $x = 1$, $$f''(x) = 6(2 \cdot 1 - 7) = -30, \text{ which is } negative.$$

Thus, $f(x)$ is the *maximum* for $x = 1$.

For $x = 6$, $$f''(x) = 6(2 \cdot 6 - 7) = 30, \text{ which is } positive.$$

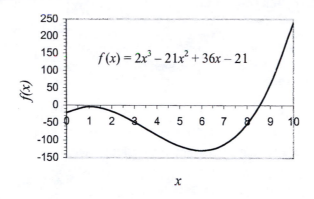

Thus, $f(x)$ is minimum for $x = 6$.

To determine the maximum and minimum values of the function, substitute the values of $x = 1$ and $x = 6$, respectively.

The maximum value of $f(x) = f(1) = 2(1^3) - 21(1^2) + 36(1) - 21 = -4$.

The minimum value of $f(x) = f(6) = 2(6^3) - 21(6^2) + 36(6) - 21 = -129$.

The plot of the function $f(x) = 2x^3 - 21x^2 + 36x - 21$ is shown showing that it has a maximum at $x = 1$ and minimum at $x = 6$.

Example 7.10.2
Investigate whether the function $f(x) = x^3 - 6x^2 + 18x + 10$ has a minimum or maximum.

$f(x) = x^3 - 6x^2 + 184x + 10$. The plot of $f(x)$ is shown below for $x = -10$ to 10.

Take the derivative with respect to x.

$$f'(x) = 3x^2 - 6(2x) + 18$$
$$= 3x^2 - 12x + 18$$

Factor out 3.

$$= 3(x^2 - 4x + 6)$$

Write as a perfect square plus a number.

$$= 3[(x^2 - 4x + 4) + 2]$$
$$= 3[(x - 2)^2 + 2]$$

Since $(x - 2)^2$ is always positive, $f'(x)$ is always positive and can never be zero. Therefore, $f(x)$ does not have a maximum nor a minimum.

Example 7.10.3 (from modern physics)

Most probable radius for an electron in a hydrogen atom: The electron in a normal hydrogen atom is at the ground state and has a radial probability distribution function (which gives the probability of finding an electron at a distance r from the nucleus of the atom) given by $P_r = 4\dfrac{r^2}{r_0^3} e^{-2r/r_0}$, where r_0 has the dimension of length and is called the *Bohr radius*. Determine the most probable distance r at which to find the electron in the ground state for a hydrogen atom.

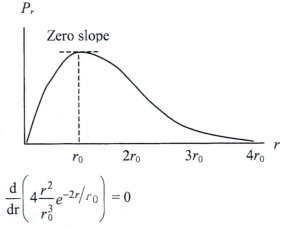

The plot of the radial probability distribution function P_r is shown at right. The peak of the curve corresponds to the most probable distance, r. At the peak, the curve has a zero slope (that is, first derivative is zero). Therefore, take the first derivative of P_r and set it equal to zero.

$$\frac{d}{dr}\left(4\frac{r^2}{r_0^3} e^{-2r/r_0}\right) = 0$$

Use the rule of derivative for the product of two functions.

Thus,

$$\left(8\frac{r}{r_0^3} - 8\frac{r^2}{r_0^4}\right) e^{-2r/r_0} = 0.$$

Divide by e^{-2r/r_0}, since it is not equal to zero except when $r = \infty$.

Thus,
$$8\frac{r}{r_0^3} - 8\frac{r^2}{r_0^4} = 0$$

Divide by 8.
$$\frac{r}{r_0^3} = \frac{r^2}{r_0^4}$$

Cancel the same quantities on both sides. $r = r_0$

Thus, the most probable distance for the electron is equal to the Bohr radius r_0 in a normal hydrogen atom.

Example 7.10.4 (from thermal physics)

Most probable speed: The velocity (v) distribution of a system of N molecules each of mass m at any temperature T (in kelvin) is given by Maxwell velocity distribution, which is expressed as

$$f(v) = 4\pi N \left(m/2\pi kT\right)^{3/2} v^2 e^{-mv^2/2kT}$$

where k is a constant called the Boltzmann constnat. *Most probable speed*, v_p, is the speed at which the number of molecules is a maximum. Determine the most probable speed.

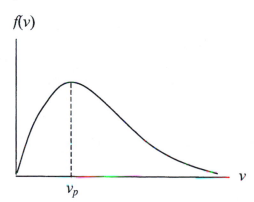

The plot of $f(v)$ is shown. For the most probable speed, $f(v)$ is a maximum,. That is, $df(v)/dv = 0$.

Thus,
$$\frac{df(v)}{dv} = \frac{d}{dv}\left(4\pi N\left(\frac{m}{2\pi kT}\right)^{3/2} v^2 e^{-mv^2/2kT}\right) = 0$$

Use the rule of products.
$$4\pi N\left(\frac{m}{2\pi kT}\right)^{3/2}\left(2v e^{-mv^2/2kT} - \frac{2mv^3}{2kT}e^{-mv^2/2kT}\right) = 0$$

Factor out $2v\,e^{-mv^2/2kT}$.
$$4\pi N 2v\,e^{-mv^2/2kT}\left(\frac{m}{2\pi kT}\right)^{3/2}\left(1 - \frac{mv^2}{2kT}\right) = 0$$

Divide by the common factor.

$$\left(1 - \frac{mv^2}{2kT}\right) = 0$$

The solution for v is the most probable speed, v_p. Therefore, $v_p = \sqrt{\dfrac{2kT}{m}}$.

Example 7.10.5 (from optics)

*Principle of least time: According to *Fermat's principle* in optics, a ray of light in passing from a point P in one medium to a point Q in another medium follows the path for which the time is a minimum. Prove the law of refraction from Fermat's principle.

Let a ray of light meet the surface of separation between the two media and POQ be a possible path of the ray. $PM (= a)$ and $QN (= b)$ are two perpendiculars drawn from P and Q to the surface. The line AOB is normal to the surface passing through the point of incidence, O. From the following figure, if v and v' are the velocities of light in the two media, the total time t taken by the ray of light is

$$t = \frac{PO}{v} + \frac{OQ}{v'} = \frac{a\sec\theta}{v} + \frac{a\sec\phi}{v'} .$$

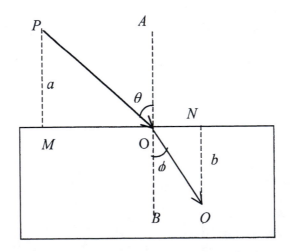

For the time t to be a minimum, set $\dfrac{dt}{d\theta} = 0$. Thus from the preceding equation for t,

$$\frac{a}{v}\sec\theta\tan\theta + \frac{b}{v'}\sec\phi\tan\phi\,\frac{d\phi}{d\theta} = 0. \tag{7.1}$$

Since the points P and Q are fixed, distance $MN = MO + ON = a \tan \theta + b \tan \phi =$ constant. Take derivative with respect to θ.

$$a \sec^2 \theta + b \sec^2 \phi \, \frac{d\phi}{d\theta} = 0 \qquad\qquad (7.2)$$

Eliminate $\dfrac{d\phi}{d\theta}$ from Equations (7.1) and (7.2). You will get after simplification (to show this determine $\dfrac{d\phi}{d\theta}$ from Equation 2 and substitute in Equation 1),

$$\frac{\sin \theta}{v} = \frac{\sin \phi}{v'}$$

$$\frac{\sin \theta}{\sin \phi} = \frac{v}{v'}.$$

This is the law of refraction.

Exercises

Find the values of x for which the following functions (Exercises 12 and 13) are maximum or minimum. Also find the maximum and minimum values of the function.

12. $f(x) = 2x^3 - 15x^2 + 36x + 9$

13. $f(x) = 2x^3 + 3x^2 - 12x + 10$

14. The force F exerted by a circular electric current of radius a on a magnet whose axis coincides with the axis of the coil is $F = kx\,(a^2 + x^2)^{-5/2}$, where k is a constant and x is the distance of the magnet from the center of the coil. Show that the force F is greatest for $x = a/2$. (*Hint*: To determine dF/dx, use the product rule or the quotient rule.)

15. The potential energy of two atoms in a diatomic molecule can be expressed by $U(r) = -a/r^6 + b/r^{12}$, where a and b are positive constants and r is the separation between the atoms. Show that $U(r)$ is minimum for $r = (2b/a)^{1/6}$. Find the value of $U(r)$ as r tends to 0 and ∞. Also, find r for which $U(r) = 0$. From these information, plot $U(r)$ as a function of r.

*16 According to Fermat's principle in optics, a ray of light in passing from a point P in one medium to a point Q in the same medium after reflection from a surface follows the path for which the time is a minimum. Prove the law of reflection (angle of incidence = angle of reflection) from Fermat's principle.

*17. The Planck's law for the radiation intensity I for a blackbody can be written as $I(x) = (2\pi k T^5/h^4 c^3) \, x^5 \, [\exp(x) - 1]^{-1}$, where $x = hc/\lambda kT$ (h, c, k are constants, λ is wavelength, and T is the temperature of the blackbody). For the wavelength for which the radiated intensity is a maximum, the derivative of $I(x)$ with respect to x is zero. Show that this leads to a transcendental equation $5e^{-x} + x - 5 = 0$.

7.11   Partial Derivatives

If there is a function of more than one independent variable, then the differentiation of the function with respect to any one of the variables is termed as *partial differentiation*. Partial differentials are common in different areas of physics especially in thermodynamics. For example, a thermodynamic system is characterized by thermodynamic variables such as pressure, volume, and temperature. All these variables cannot be changed independently; when certain ones are changed, the remaining ones are automatically fixed in value. This situation is mathematically represented by partial differentiation.

If $z = f(x, y)$, the partial derivative of $f(x, y)$ with respect to x, keeping y constant, is mathematically defined as

$$\frac{\partial f}{\partial x} = \lim_{\Delta x \to 0} \frac{f(x + \Delta x, y) - f(x, y)}{\Delta x}.$$

Similarly, the partial derivative of $f(x, y)$ with respect to y, keeping x constant, is defined as

$$\frac{\partial f}{\partial y} = \lim_{\Delta y \to 0} \frac{f(x, y + \Delta y) - f(x, y)}{\Delta y}.$$

To take the partial derivative of a function $z = f(x, y, ...,)$ with respect to x, first pretend all variables, except the first variable x, are constant and take the derivative of the function with respect to x. This is $\dfrac{\partial f}{\partial x}$. Next, pretend all variables except the second variable y are constant and take the derivative with respect to y. This is $\dfrac{\partial f}{\partial y}$.

Repeat this process for any other variables in the function.

Note: The curly ∂ is commonly used to denote the symbol of partial derivative, to distinguish it from the symbol d of ordinary derivative.

Example 7.11.1

Determine $\dfrac{\partial f}{\partial x}$ or $\dfrac{\partial z}{\partial x}$ and $\dfrac{\partial f}{\partial y}$ or $\dfrac{\partial z}{\partial y}$ for the function $z = f(x, y) = x^2 + 2xy + y^3$.

The partial derivative of the function with respect to x,

$$\frac{\partial f}{\partial x} \text{ or } \frac{\partial z}{\partial x} = \frac{\partial}{\partial x}(x^2 + 2xy + y^3)$$

Take the derivative, assuming y is constant.

$$= 2x + 2y \cdot 1 + 0$$
$$= 2(x + y).$$

The partial derivative of $f(x, y)$ with respect to y,

$$\frac{\partial f}{\partial y} \text{ or } \frac{\partial z}{\partial y} = \frac{\partial}{\partial y}(x^2 + 2xy + y^3)$$

Take the derivative, assuming x constant.

$$= 0 + 2x \cdot 1 + 3y^2$$
$$= 2x + 3y^2$$

The partial derivatives of second order are defined as follows:

$$\frac{\partial^2 f}{\partial x^2} = \frac{\partial}{\partial x}\left(\frac{\partial f}{\partial x}\right) \qquad\qquad \frac{\partial^2 f}{\partial y^2} = \frac{\partial}{\partial y}\left(\frac{\partial f}{\partial y}\right)$$

$$\frac{\partial^2 f}{\partial x\, \partial y} = \frac{\partial}{\partial x}\left(\frac{\partial f}{\partial y}\right) \qquad\qquad \frac{\partial^2 f}{\partial y\, \partial x} = \frac{\partial}{\partial y}\left(\frac{\partial f}{\partial x}\right).$$

The results of differentiation from the bottom two expressions will be the same if the function and its partial derivative are continuous. In this case, the order of differentiation is not important. Similarly, other higher-order partial derivates are defined.

Example 7.11.2

A wave propagating to the left can be expreseed by the function $y(x, t) = A\sin(kx + \omega t)$ where A, k and ω are constants, called the amplitude, wave number, and the angular frequency of the wave, respectively. Show that the wave satisfies the equation $\partial^2 y/\partial t^2 = v^2\, \partial^2 y/\partial x^2$, where $v = \omega/k$.

$y(x, t) = A\sin(kx + \omega t)$. Here the variables are x and t.
First, take the partial derivative of the equation twice with respect to t: Use the results for the derivatives of sin and cos functions from Section 7.5.

$$\frac{\partial y}{\partial t} = \frac{\partial}{\partial t} [A \sin(kx + \omega t)] = A \omega \cos(kx + \omega t)$$

$$\frac{\partial^2 y}{\partial t^2} = \frac{\partial}{\partial t} [A \omega \cos(kx + \omega t)] = A\omega [-\omega \sin(kx + \omega t)] = -A\omega^2 \sin(kx + \omega t)$$

Next, take the partial derivative of the equation twice with respect to x.

$$\frac{\partial y}{\partial x} = \frac{\partial}{\partial x} [A \sin(kx + \omega t)] = A k \cos(kx + \omega t)$$

$$\frac{\partial^2 y}{\partial x^2} = \frac{\partial}{\partial x} [A k \cos(kx + \omega t)] = Ak [-k \sin(kx + \omega t)] = -Ak^2 \sin(kx + \omega t)$$

Thus, dividing the two second derivatives, we get

$$\frac{\partial^2 y}{\partial t^2} \bigg/ \frac{\partial^2 y}{\partial x^2} = \frac{-A\omega^2 \sin(kx - \omega t)}{-Ak^2 \sin(kx - \omega t)} = \frac{\omega^2}{k^2} = v^2 \text{ (since } v = \omega/k)$$

or $$\frac{\partial^2 y}{\partial t^2} = v^2 \frac{\partial^2 y}{\partial x^2}.$$

This equation is called the wave equation in physics.

Exercises

18. Show that $\dfrac{\partial^2 z}{\partial x^2} + \dfrac{\partial^2 z}{\partial y^2} = 0$, if $z = \log(x^2 + y^2)$.

19. If $f(x, y) = x^3 + x^2y^2 + y^3$, show that $\dfrac{\partial^2 f}{\partial x \, \partial y} = \dfrac{\partial^2 f}{\partial y \, \partial x}$.

20. If $y = f(x + vt) + \phi(x - vt)$, show that $\dfrac{\partial^2 y}{\partial t^2} = v^2 \dfrac{\partial^2 y}{\partial x^2}$.

21. The relation between pressure P, volume V and temperature T of n mols of a perfect gas is $PV = nRT$ where R is a constant. Show that $(dP/dV)_T = -nRT/V^2$ and $(dP/dT)_V = nR/V$, where $(dP/dV)_T$ means the partial derivative of P with respect to V at constant T, etc.

*7.12 Exact or Perfect Differential

The *differential dz* of the function $z = f(x, y)$ is defined as

$$dz = \left(\frac{\partial z}{\partial x}\right)_y dx + \left(\frac{\partial z}{\partial y}\right)_x dy.$$

Here, the subscript x (or y) means that the derivative was taken when x (or y) was constant.

The expression $dz = M(x, y)\, dx + N(x, y)\, dy$ is said to an *exact or perfect differential* if a function z of x, y exists such that $\left(\dfrac{\partial M}{\partial y}\right)_x = \left(\dfrac{\partial N}{\partial x}\right)_y$.

Taking derivatives of the preceding expressions with respect to y and x, respectively,

$$\frac{\partial^2 z}{\partial y\, \partial x} = \left(\frac{\partial M}{\partial y}\right)_x \qquad \text{and} \qquad \frac{\partial^2 z}{\partial x\, \partial y} = \left(\frac{\partial N}{\partial x}\right)_y.$$

Thus, the condition for perfect differential is

$$\left(\frac{\partial M}{\partial y}\right)_x = \left(\frac{\partial N}{\partial x}\right)_y \qquad \text{or} \qquad \frac{\partial^2 z}{\partial y\, \partial x} = \frac{\partial^2 z}{\partial x\, \partial y}.$$

*Example 7.12.1 (from thermodynamics)

The four thermodynamic functions called internal energy (U), enthalpy (H), free energy (F), and Gibbs's potential (G) are perfect differentials. The differentials of these functions are expressed as, respectively,

$$dU = T\, dS - P\, dV,$$

$$dH = T\, dS + V\, dP,$$

$$dF = -S\, dT - P\, dV,$$

$$dG = -S\, dT + V\, dP.$$

Here, P, V, T, and S are other thermodynamic variables called pressure, volume, temperature, and entropy, respectively. Express the condition for each of these perfect differentials.

To solve this problem, compare the differentials with the expression $M\, dx + N\, dy$ and write down condition as $\left(\dfrac{\partial M}{\partial y}\right)_x = \left(\dfrac{\partial N}{\partial x}\right)_y.$

Thus, from the expression for dU,
$$\left(\frac{\partial T}{\partial V}\right)_S = -\left(\frac{\partial P}{\partial S}\right)_V.$$

From the expression for dH,
$$\left(\frac{\partial T}{\partial P}\right)_S = \left(\frac{\partial V}{\partial S}\right)_P.$$

From the expression for dF,
$$\left(\frac{\partial S}{\partial V}\right)_T = \left(\frac{\partial P}{\partial T}\right)_V.$$

From the expression for dG,
$$\left(\frac{\partial S}{\partial P}\right)_T = -\left(\frac{\partial V}{\partial T}\right)_P.$$

These four relations are called *Maxwell's four thermodynamic relations*.

*Example 7.12.2 (from thermodynamics)

The relationship between pressure P volume V and temperature T of a thermodynamic system can be expressed by the function $f(P, V, T) = 0$. The equation can be written as $T = T(P, V)$ (that is, T is a function of P and V) and also as $P = P(T, V)$. Show that (i)

(i) $\left(\frac{\partial P}{\partial T}\right)_V = 1 \Big/ \left(\frac{\partial T}{\partial P}\right)_V$ and (ii) $\left(\frac{\partial P}{\partial T}\right)_V \left(\frac{\partial T}{\partial V}\right)_P \left(\frac{\partial V}{\partial P}\right)_T = -1.$

From the function $T = T(P, V)$,
$$dT = \left(\frac{\partial T}{\partial P}\right)_V dP + \left(\frac{\partial T}{\partial V}\right)_P dV.$$

From the function $P = P(T, V)$,
$$dP = \left(\frac{\partial P}{\partial T}\right)_V dT + \left(\frac{\partial P}{\partial V}\right)_T dV.$$

Substituting the value of dT,

$$dP = \left(\frac{\partial P}{\partial T}\right)_V \left[\left(\frac{\partial T}{\partial P}\right)_V dP + \left(\frac{\partial T}{\partial V}\right)_P dV\right] + \left(\frac{\partial P}{\partial V}\right)_T dV$$

$$= \left(\frac{\partial P}{\partial T}\right)_V \left(\frac{\partial T}{\partial P}\right)_V dP + \left[\left(\frac{\partial P}{\partial T}\right)_V \left(\frac{\partial T}{\partial V}\right)_P + \left(\frac{\partial P}{\partial V}\right)_T\right] dV.$$

Transferring all terms to the left-hand side,

$$\left[1 - \left(\frac{\partial P}{\partial T}\right)_V \left(\frac{\partial T}{\partial P}\right)_V\right] dP - \left[\left(\frac{\partial P}{\partial T}\right)_V \left(\frac{\partial T}{\partial V}\right)_P + \left(\frac{\partial P}{\partial V}\right)_T\right] dV = 0$$

Since P and V can be independent variables, the left-hand side can be zero only if the coefficients of dP and dV are zero; that is,

(i) $1 - \left(\dfrac{\partial P}{\partial T}\right)_V \left(\dfrac{\partial T}{\partial P}\right)_V = 0.$

That is, $\left(\dfrac{\partial P}{\partial T}\right)_V \left(\dfrac{\partial T}{\partial P}\right)_V = 1$ or $\left(\dfrac{\partial P}{\partial T}\right)_V = 1 \bigg/ \left(\dfrac{\partial T}{\partial P}\right)_V$

(ii) $\left(\dfrac{\partial P}{\partial T}\right)_V \left(\dfrac{\partial T}{\partial V}\right)_P + \left(\dfrac{\partial P}{\partial V}\right)_T = 0$

$\left(\dfrac{\partial P}{\partial T}\right)_V \left(\dfrac{\partial T}{\partial V}\right)_P = -\left(\dfrac{\partial P}{\partial V}\right)_T = -1 \bigg/ \left(\dfrac{\partial V}{\partial P}\right)_T$ or $\left(\dfrac{\partial P}{\partial T}\right)_V \left(\dfrac{\partial T}{\partial V}\right)_P \left(\dfrac{\partial V}{\partial P}\right)_T = -1$

Extension of partial differentiation and differential to functions of more than two variables are exactly analogous.

Exercise
22. Show that

(a) $\left(\dfrac{\partial V}{\partial T}\right)_P = -\left(\dfrac{\partial P}{\partial T}\right)_V \bigg/ \left(\dfrac{\partial P}{\partial V}\right)_T$ (b) $\left(\dfrac{\partial P}{\partial T}\right)_V = -\left(\dfrac{\partial V}{\partial T}\right)_P \bigg/ \left(\dfrac{\partial V}{\partial P}\right)_T$

*7.13 Partial Derivatives of a Function of Two Functions

Let $z = f(x_1, x_2)$, where $x_1 = f_1(x, y)$ and $x_2 = f_2(x, y)$ and x, y are independent variables. Then

$$\frac{\partial z}{\partial x} = \frac{\partial z}{\partial x_1} \frac{\partial x_1}{\partial x} + \frac{\partial z}{\partial x_2} \frac{\partial x_2}{\partial x}$$

$$\frac{\partial z}{\partial y} = \frac{\partial z}{\partial x_1} \frac{\partial x_1}{\partial y} + \frac{\partial z}{\partial x_2} \frac{\partial x_2}{\partial y}.$$

The result can be generalized to the cases of more than two variables.

Answers to Exercises in Chapter 7

1. $21x^6 + 16x^3 - 4x - 1$

2. $x + 2x \log x$

3. $(x^2 + 1) \cos x + 2x \sin x$

4. $(x \cos x - \sin x)/x^2$

5. $\dfrac{2\cos x}{\left(1 - \sin x\right)^2}$

6. $-(b/a)\cot t$

7. $-\tan \dfrac{1}{2} t$

8. $\dfrac{2x(6x^2 - y)}{x^2 - 6y^2}$

9. $-\left(\dfrac{y}{x}\right)^{\frac{1}{3}}$

10. $v = (2.8 + 8.4\, t)$ m/s, $a = 8.4$ m/s^2

12. $f(x)$ is maximum for $x = 2$ and minimum for $x = 3$; $f_{max} = f(2) = 37$ and $f_{min} = f(3) = 36$.

13. $f(x)$ is maximum for $x = -2$ and minimum for $x = 1$; $f_{max} = f(-2) = 30$ and $f_{min} = f(1) = 3$.

15. At $r \to 0$, $U(r) \to \infty$; as $r \to \infty$, $U(r) \to 0$, $U(r) = 0$ for $r = (b/a)^{1/6}$.

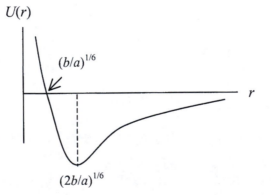

Chapter 8
Integral Calculus and Differential Equations

Integral Calculus

Indefinite Integrals

8.1 Integration: A Way to Find Total

The word *integration* literally means "summation." It is the process of determining the limit of a sum of a certain number of elements as the number of elements tends to infinity and each element becomes infinitesimally small. The technique is applied to a wide variety of geometrical and physical problems such as finding arc length, volume, moment of inertia, work, fluid pressure and so on. It has played an important role in the development of physics.

Integration can also be viewed as the *inverse of differentiation*. The determination of derivative or differential expression raises the reverse problem of passing from derivative or differential expression to the original, called the primitive. The equivalence of both viewpoints of integration has been established.

Integration defined as the inverse of a derivative or of a differential expression can be presented in the form of a question: What is the original function that when differentiated produces a given function? The process of finding the function whose derivative or differential is known is called *integration*. For example, what is the function that when differentiated produces $3x^2$? You know from your knowledge of differentiation that the function is x^3. Thus, x^3 is an integral of $3x^2$ with respect to x.

In general, if $\dfrac{d}{dx} f(x) = \phi(x)$ or $d f(x) = \phi(x)\, dx$, then $f(x)$ is the integral

of $\phi(x)$ with respect to x. This is written mathematically as $\int \phi(x)\, dx = f(x)$.

Constant of Integration

Note that $d f(x) = \phi(x)\, dx$. That is, $\int \phi(x)\, dx = f(x)$.

Also, $d[f(x) + c] = \phi(x)\, dx$,   since $d(c) = 0$ (differential of a constant is zero).

That is, $\int \phi(x)\, dx = f(x) + c$.

Here c is an arbitrary constant independent of the variable of integration. Thus we find that *an arbitrary constant must be added* to get a complete integral. $\int \phi(x)\, dx$ is thus called an *indefinite integral* because the value of the integral is not definite unless c is known. In practical physical problems, you will be given such conditions to determine c, and in such situations definite solution of the integral will be obtained.

Theorems of Integration

1. *The integral of the sum or difference of a finite number of functions is equal to the sum or difference of their integrals.* That is,

$$\int [\phi_1 \pm \phi_2 \pm \phi_3 \pm \cdots \pm \phi_n)\, dx = \int \phi_1\, dx \pm \int \phi_2\, dx \pm \int \phi_3\, dx \pm \cdots \pm \int \phi_n\, dx.$$

2. *A constant factor can be taken out from under the sign of integration and placed before it.* That is, $\int A\phi(x)\, dx = A\int \phi(x)\, dx$, where A is a constant.

8.2   Table of Indefinite Integrals of Common Functions

The knowledge of differentiation is employed to determine indefinite integrals of different functions. Here is the list of the results of some common indefinite integrals that are useful in physics. The arbitrary constant of integration in all cases needs to be added. Like derivatives, determining integrals in physical problems is mostly a matter of choosing an integral from an appropriate list, such as given below, that has the same form as the one you find in the problem.

$$\int x^n\, dx = \frac{x^{n+1}}{n+1} \qquad\qquad \int (a+bx)^n\, dx = \frac{(a+bx)^{n+1}}{b(n+1)}$$

$$\int e^{mx}\, dx = \frac{e^{mx}}{m} \qquad\qquad \int \frac{dx}{x} = \ln x$$

$$\int \sin mx\, dx = -\frac{\cos mx}{m} \qquad\qquad \int \cos mx\, dx = \frac{\sin mx}{m}$$

$$\int \tan mx\, dx = \frac{1}{m}\ln\left|\sec mx\right| \qquad\qquad \int \cot mx\, dx = \frac{1}{m}\ln\left|\sin mx\right|$$

$$\int \sec mx\, dx = \frac{1}{m}\ln\left|\sec mx + \tan mx\right| \qquad \int \csc mx\, dx = \frac{1}{m}\ln\left|\csc mx - \cot mx\right|$$

$$\int \sec^2 mx\, dx = \frac{\tan mx}{m} \qquad\qquad \int \csc^2 x\, dx = -\frac{\cot mx}{m}$$

$$\int \sin^2 x\, dx = \frac{1}{2}x - \frac{1}{4}\sin 2x \qquad\qquad \int \cos^2 x\, dx = \frac{1}{2}x + \frac{1}{4}\sin 2x$$

$$\int x \sin^2 x \, dx = \frac{1}{4}x^2 - \frac{1}{4}x \sin 2x - \frac{1}{8}\cos 2x$$

$$\int x^2 \sin^2 x \, dx = \frac{1}{6}x^3 - \left(\frac{x^2}{4} - \frac{1}{8}\right)\sin 2x - \frac{1}{4}x \cos 2x$$

$$\int \frac{dx}{a^2 + x^2} = \frac{1}{a}\tan^{-1}\frac{x}{a} \qquad\qquad \int \frac{x \, dx}{x^2 + a^2} = \frac{1}{2}\ln(x^2 + a^2)$$

$$\int \frac{dx}{x^2 - a^2} = \frac{1}{2a}\log\left|\frac{x-a}{x+a}\right| \qquad\qquad \int \frac{dx}{\sqrt{x^2 \pm a^2}} = \ln\left(x + \sqrt{x^2 \pm a^2}\right)$$

$$\int \frac{dx}{\sqrt{a^2 - x^2}} = \sin^{-1}\frac{x}{a} \qquad\qquad \int \frac{x \, dx}{\sqrt{x^2 \pm a^2}} = \sqrt{x^2 \pm a^2}$$

$$\int \frac{dx}{\left(x^2 \pm a^2\right)^{3/2}} = \frac{1}{a^2}\left(\frac{\pm x}{\sqrt{x^2 \pm a^2}}\right) \qquad\qquad \int \frac{x \, dx}{\left(x^2 + a^2\right)^{3/2}} = -\frac{1}{\sqrt{x^2 + a^2}}$$

$$\int \frac{dx}{a + bx} = \frac{1}{b}\ln(a + bx) \qquad\qquad \int \frac{dx}{(a + bx)^2} = -\frac{1}{b(a + bx)}$$

$$\int \frac{x \, dx}{a + bx^2} = \frac{1}{2b}\ln(a + bx^2)$$

$$\int \frac{dx}{a + bx^2} = \frac{1}{\sqrt{ab}}\tan^{-1}\left(\frac{\sqrt{b}}{a}x\right)(\text{if } ab > 0); \quad \frac{1}{2\sqrt{ab}}\ln\left(\frac{a - x\sqrt{|ab|}}{a + x\sqrt{|ab|}}\right)(\text{if } ab < 0)$$

$$\int xe^{mx} \, dx = \frac{e^{mx}}{m^2}(mx - 1)$$

It is not a good idea to depend on the list for simple integrals. You should *memorize* the following results:

The integral of zero is an arbitrary constant c.

The integral of x^n is x^{n+1}/n.

The integral of e^{mx} is e^{mx}/m.

The integral of $1/x$ is $\ln x$.

The integral of $\sin mx$ is $-\cos mx/m$.

The integral of $\cos mx$ is $\sin mx/m$.

The integral of $\tan mx$ is $\ln|\sec mx|/m$.

Also, note that some calculators can do integrals without any need of integral tables. You will get more information from the owner's manual.

Some integrals may have more complex functions than listed in the integration tables. But many of these functions can be rewritten so that the table will be useful for evaluating the integrals. The following techniques are helpful.

8.3   Transformations of Simple Trigonometric Functions

Certain trigonometric identities are useful for determining integrals and are frequently used to change powers or products of trigonometric functions into algebraic sum or difference of other functions or in terms of multiple angles.

Power of Trigonometric Functions

$$\sin^2 x = \frac{1}{2}(1 - \cos 2x) \qquad\qquad \cos^2 x = \frac{1}{2}(1 + \cos 2x)$$

$$\tan^2 x = \sec^2 x - 1 \qquad\qquad \cot^2 x = \csc^2 x - 1$$

$$\sin^3 x = \frac{1}{4}(3\sin x - \sin 3x) \qquad \cos^3 x = \frac{1}{4}(3\cos x + \cos 3x)$$

Example 8.3.1

Evaluate the integral $\int \sin^3 x \, dx$.

$$\int \sin^3 x \, dx = \int \frac{1}{4}(3\sin x - \sin 3x)\, dx = \frac{1}{4}\left[3\int \sin x \, dx - \int \sin 3x \, dx\right]$$

Use the result from the table.
$$= \frac{1}{4}\left[3(-\cos x) - \frac{1}{3}(-\cos 3x)\right] + c$$

$$= \frac{1}{4}\left[-3\cos x + \frac{1}{3}\cos 3x)\right] + c$$

Products of sin and cos, Integrals of the Form [$m \neq n$]
$$\int \sin mx \cos nx \, dx, \quad \int \sin mx \sin nx \, dx, \text{ and } \int \cos mx \cos nx \, dx$$

Use the following trigonometric identities:

$$\sin A \cos B = \frac{1}{2}\sin(A + B) + \frac{1}{2}\sin(A - B).$$

$$\cos A \cos B = \frac{1}{2}\cos(A - B) + \frac{1}{2}\cos(A + B).$$

$$\sin A \sin B = \frac{1}{2}\cos(A - B) - \frac{1}{2}\cos(A + B).$$

Other trigonometric identities are also often useful in evaluating an integral involving trigonometric functions.

Example 8.3.2

Evaluate the integral $\int \sin 2x \cos 4x \, dx$.

$$\int \sin 2x \cos 4x \, dx$$

Use the trigonometric identity.
$$= \int [\frac{1}{2}\sin(4x + 2x) + \frac{1}{2}\sin(4x - 2x)] \, dx$$

$$= \frac{1}{2}\int \sin 6x \, dx + \frac{1}{2}\int \sin 2x \, dx$$

Use the result from the integral table.
$$= \frac{1}{2}\left(-\frac{\cos 6x}{6}\right) + \frac{1}{2}\left(-\frac{\cos 2x}{2}\right) + c$$

$$= -\frac{1}{4}\left[\frac{\cos 6x}{3} + \cos 2x\right] + c$$

Exercises

Evaluate the following integrals:

1. $\int \sin^2 x \, dx$

2. $\int \sin mx \sin nx \, dx.$

8.4   Change of Variable or Method of Substitution

In many situations changing variables in the integrand helps to match the integral with a standard integral listed in the integral table (Section 7.15). Carefully choose a term in the integrand to simplify the problem and set the expression to a new variable, say, u. Determine du in terms of dx and rewrite the integral in terms of the new variable u. Evaluate the integral (which is a function of u). Finally, plug in the values of u in the result in terms of x. (For definite integrals [discussed later], determine the upper and lower limits of the new variable u, these limits are the new limits of integration.)

Example 8.4.1

Evaluate the integral $\int \sin x \cos x \, dx$.

This is not a standard integral that usually is listed in an integral table. However, by carefully choosing a new variable, the integral can be converted to a standard form. Let $\sin x = u$. Then $\cos x\, dx = du$. Therefore,

$$\int \sin x \cos x\, dx = \int \sin x\, (\cos x\, dx)$$

Substitute u for $\sin x$ and du for $\cos x\, dx$.

$$= \int u\, du$$

Use the result from the integration table.

$$= \frac{u^2}{2} + c$$

Substitute $\sin x$ for u.

$$= \frac{\sin^2 x}{2} + c.$$

Example 8.4.2

Evaluate the integral $\displaystyle\int \frac{\sin^{-1} x\, dx}{\sqrt{1 - x^2}}$.

This is also not a standard integral that usually is listed in an integral table. However, as shown below, the integral can be easily converted to a standard form.

Let $\sin^{-1} x = u$, that is, $\dfrac{1}{\sqrt{1 - x^2}}\, dx = du$. Therefore,

$$\int \frac{\sin^{-1} x\, dx}{\sqrt{1 - x^2}} = \int u\, du = \frac{u^2}{2} + c = \frac{\left(\sin^{-1} x\right)^2}{2} + c.$$

Exercises

Evaluate the following integrals by the method of substitution:

3. $\displaystyle\int \frac{\log x\, dx}{x}$

4. $\displaystyle\int \frac{\cos x\, dx}{\sin^2 x + 1}$

5. $\displaystyle\int \frac{dx}{\sqrt{9 - 4x^2}}$

6. $\displaystyle\int \frac{e^{2x}\, dx}{e^x + 1}$

8.5   Rules of Integration

Integration by Parts

If u and v are two differentiable functions of a single variable, then

$$\int u \, dv = uv - \int v \, du.$$

This rule of integration is known as *integration by parts*, which is very useful in evaluating thorny integrals.

To use this method, you must rightly choose u and dv so that the product $u \, dv$ is the expression to be integrated. There is no rule for deciding how to divide the integrand into u and dv, but a good rule of thumb is to try the more complicated term in the integrand as dv first.

Example 8.5.1

Evaluate the integral $\int x \, e^x \, dx.$

Let $u = x$ and $dv = e^x \, dx.$

Then $du = dx$ and $v = e^x.$

Therefore, $\int x \, e^x \, dx. = \int u \, dv.$

Thus, using the rule of integrating by parts,

$$\int x \, e^x \, dx. = x \, e^x - \int e^x \, dx.$$

$$= x \, e^x - e^x + c$$

$$= x(e^x - 1) + c.$$

Exercises

Evaluate the following integrals by the method of integration by parts:

7. $\int x \, \cos x \, dx$

8. $\int \log x \, dx.$

Definite Integrals

8.6   Definite Integral and the Fundamental Theorem of Calculus; Areas and Definite Integrals

Definite integrals specify the endpoints of the integral so that the result of the integral is definite or exact. The symbol $\int_a^b \phi(x)\, dx$ represents the definite integral of the function $\phi(x)$ with respect to x as changes from a to b. a and b are called the limits of the integral, a being the lower limit and b the upper limit. The interval (a, b) is called the range of the integral.

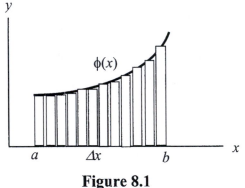

Figure 8.1

Let the function $\phi(x)$ be defined in an interval $a \leq x \leq b$. Let us divide the interval between a and b into n equal parts. The length of the interval is then $(b - a)$, and, therefore, the length Δx of each part is $\Delta x = (b - a)/n$ (Figure 8.1). Then the definite integral of $\phi(x)$ between a and b is defined as the limit

$$\int_a^b \phi(x)\, dx$$

$$= \lim_{n \to \infty} [\phi(a)\, \Delta x + \phi(a + \Delta x)\, \Delta x + \phi(a + 2\, \Delta x)\, \Delta x + \cdots + \phi(a + (n - 1)\, \Delta x)\, \Delta x].$$

This limit will exist if $\phi(x)$ is piecewise continuous.

Geometrically, the product $\phi(a)\, \Delta x$ represents the area of a rectangle of height $\phi(a)$ and width Δx. Thus, the summation of product of each term represents the total area of all the rectangles as shown in Figure 8.1. In the limit n tends to infinity the width Δx becomes very small and the summation of the areas of all the rectangles becomes practically the area between the curve and the x-axis between the limits $x = a$ and $x = b$.

Definite integrals are calculated via the *fundamental theorem of calculus*:

$$\int_a^b \phi(x)\, dx = f(x)\Big|_a^b = f(b) - f(a).$$

Note: To evaluate a definite integral such as $\int_a^b \phi(x)\,dx$,

Determine the indefinite integral $\int \phi(x)\,dx$, the result is, say, $f(x)$.

Substitute the upper limit, $x = b$, in $f(x)$ to determine $f(b)$.

Substitute the lower limit, $x = a$, in $f(x)$ to determine $f(a)$.

Subtract $f(a)$ from $f(b)$ to determine the result of the integral.

Example 8.6.1

Determine the following definite integrals:

(a) $\displaystyle\int_0^2 x^4\,dx$

(b) $\displaystyle\int_0^{1/2} \frac{dx}{\sqrt{1-x^2}}$

(a) To evaluate the definite integral $\displaystyle\int_0^2 x^4\,dx$, first determine the indefinite integral

$\displaystyle\int x^4\,dx$ using the table of integration and then use the fundamental theorem of

calculus.

$$\int x^4\,dx = \frac{x^{4+1}}{4+1} = \frac{x^5}{5}$$

Therefore,

$$\int_0^2 x^4\,dx = \left.\frac{x^5}{5}\right|_0^2 = \frac{2^5}{5} - \frac{0^5}{5} = \frac{32}{5} = 6\frac{2}{5}.$$

(b) From the table of indefinite integral, $\displaystyle\int \frac{dx}{\sqrt{1-x^2}} = \sin^{-1}x.$

Therefore,

$$\int_{0}^{1/2} \frac{dx}{\sqrt{1-x^2}} = \left. \sin^{-1} x \right|_{0}^{1/2} = \sin^{-1} \frac{1}{2} - \sin^{-1}(0) = \frac{\pi}{6} - 0 = \frac{\pi}{6}.$$

Example 8.6.2 (from mechanics)

The force, F, needed to stretch (or compress) a linear spring is proportional to the amount of stretch (or compression) x; that is $F = kx$. Determine the total work done, W, to stretch a spring from $x = 0$ to $x = x_0$ if work is defined as work = force X distance.

The work done to stretch the spring by an infinitesimal amount dx = force X distance = $F\,dx = kx\,dx$.

Therefore, the total work done to stretch the spring from $x = 0$ to $x = x$ is given by the definite integral

$$W = \int_{x=0}^{x\,=x_0} kx\,dx = k \int_{x=0}^{x\,=\,x_0} x\,dx$$

Use fundamental theorem of calculus.
$$= \left. k \frac{x^2}{2} \right|_{0}^{x_0} = k[\frac{x_0^2}{2} - \frac{0^2}{2}] = \frac{1}{2} kx_0^2.$$

Example 8.6.2 (from thermal physics)

Entropy of a substance is mesurement of its molecular disorder. When a substance (such as water) is heated from temperature (in Kelvin, K) T_1 to T_2, its increase in entropy is given by $ds = \int_{T_1}^{T_2} mc \frac{dT}{T}$, where m is the mass and c is a constant (called the specific heat) of the substance. Determine the increase in entropy of 1.00 kg of water when it is heated from 273 K to 373 K. For water, $c = 4186$ J/kg · K.

In this case, $m = 1.00$ kg, $c = 4186$ J/(kg · K), $T_1 = 273$ K and $T_2 = 373$ K.

$$ds = \int_{T_1}^{T_2} mc \frac{dT}{T}$$

Take m, c out of the integral.
$$= mc \int_{T_1}^{T_2} \frac{dT}{T}$$

Substitute the values.

$$= (1.00 \text{ kg}) (4186 \text{ J/kg·K}) \int_{273}^{373} \frac{dT}{T}$$

Evaluate the integral.

$$= (4186 \text{ J/K}) \ln\left(\frac{373}{273}\right) = 1306 \text{ J/K}$$

Example 8.6.4 (from modern physics)

The average value of the location of a particle in an infinite well of width a is given by

$\langle x \rangle = \int_0^a x |\Psi|^2 \, dx$, where $\Psi = \sqrt{\dfrac{2}{a}} \sin \dfrac{n\pi x}{a}$. Show that $\langle x \rangle = a/2$.

The average value of the location of a particle in an infinite well of width a is given by

$$\langle x \rangle \quad = \int_0^a x |\Psi|^2 \, dx$$

Substitute $\Psi = \sqrt{\dfrac{2}{a}} \sin \dfrac{n\pi x}{a}$.

$$= \int_0^a x \left(\frac{2}{a} \sin^2 \frac{n\pi x}{a} \right) dx$$

Take the factor $\dfrac{2}{a}$ out of the integral.

$$= \frac{2}{a} \int_0^a x \, \sin^2 \frac{n\pi x}{a} \, dx .$$

Change the variable $\theta = n\pi x/a$ so that $d\theta = (n\pi/a) \, dx$.

Note that when $x = 0$, $\theta = 0$ and when $x = a$, $\theta = n\pi$. Thus,

$$\langle x \rangle = \frac{2}{a} \int_0^{n\pi} \left(\frac{a}{n\pi} \right) \theta \, \sin^2 \theta \left(\frac{a}{n\pi} \right) d\theta$$

Take out the constant factors.

$$= \frac{2}{a} \left(\frac{a}{n\pi} \right)^2 \int_0^{n\pi} \theta \, \sin^2 \theta \, d\theta .$$

Use the result of the integral $\int \theta \sin^2\theta \, d\theta$ from Section 8.2 and put in the limits.

$$\langle x \rangle = \frac{2}{a}\left(\frac{a}{n\pi}\right)^2 \left[\frac{1}{4}\theta^2 - \frac{1}{4}\theta\sin 2\theta - \frac{1}{8}\cos 2\theta + c\right]_0^{n\pi}$$

$$= \frac{2a}{n^2\pi^2}\left[\frac{n^2\pi^2}{4} - 0 - \frac{1}{8} + \frac{1}{8}\right] = \frac{a}{2}$$

8.7   Common Definite Integrals

Some common definite integrals that appear in physics are listed below.

$$\int_0^\infty x^n e^{-ax}\,dx = \frac{n!}{a^{n+1}} \qquad \int_0^\infty x^{2n} e^{-ax^2}\,dx = \frac{(2n-1)!}{2^{n+1}a^n}\sqrt{\frac{\pi}{a}}$$

$$\int_0^\infty e^{-ax^2}\,dx = \frac{1}{2}\sqrt{\frac{\pi}{a}} \qquad \int_0^\infty \frac{\sin mx}{x}\,dx = \frac{\pi}{2} \text{ if } m > 0;\ 0 \text{ if } m = 0;\ -\frac{\pi}{2} \text{ if } m < 0$$

$$\int_0^\infty \frac{dx}{\left(1+e^{ax}\right)}\,dx = \frac{\pi}{2} \qquad \int_0^a \frac{dx}{\sqrt{a^2-x^2}} = \frac{\pi}{2}$$

$$\int_0^a \sqrt{a^2-x^2}\,dx = \frac{\pi a^2}{4} \qquad \int_0^\infty \frac{dx}{x^2+a^2} = \frac{\pi}{2a}$$

$$\int_0^\infty \frac{x^3\,dx}{e^x - 1} = \frac{\pi^4}{15} \qquad \int_0^{\frac{\pi}{2}} \sin^2 x\,dx = \int_0^{\frac{\pi}{2}} \cos^2 x\,dx = \frac{\pi}{4}$$

$$\int_0^\pi \sin mx \sin nx\,dx = 0 \text{ (if } m, n \text{ integers and } m \neq n\text{)}, \ \frac{\pi}{2} \text{ (if } m, n \text{ integers and } m = n\text{)}$$

$$\int_0^\pi \cos mx \cos nx\,dx = 0 \text{ (if } m, n \text{ integers and } m \neq n\text{)}, \ \frac{\pi}{2} \text{ (if } m, n \text{ integers and } m = n\text{)}$$

$$\int_0^\pi \sin mx \cos nx\,dx = 0 \text{ (if } m, n \text{ integers and } m + n \text{ odd}\text{)}, \ 2m/(m^2 - n^2) \text{ (if } m, n \text{ integers}$$

and $m + n$ even)

$$\int_0^{2\pi} \sin mx \, \sin nx \, dx = 0 \text{ (if } m, n \text{ integers and } m \neq n), \ \pi \text{ (if } m, n \text{ integers and } m = n)$$

$$\int_0^{2\pi} \cos mx \, \cos nx \, dx = 0 \text{ (if } m, n \text{ integers and } m \neq n), \ \pi \text{ (if } m, n \text{ integers and } m = n)$$

$$\int_0^{2\pi} \sin mx \, \cos nx \, dx = 0 \text{ (if } m, n \text{ integers)}$$

A good list of integrals (indefinite and definite) is extensive. You should look at the *CRC Handbook of Chemistry and Physics* (CRC Press, New York) or other mathematics book that contain tables of integrals.

Exercises
Show that

9. $\displaystyle\int_0^{\frac{\pi}{2}} \sin^2 \theta \, d\theta = \frac{\pi}{4}$

10. $\displaystyle\int_0^{\frac{\pi}{2}} \cos \theta \cos 3\theta \, d\theta = 0$

Evaluate the following definite integrals:

11. $\displaystyle\int_0^{\frac{\pi}{2}} \sin^5 x \, \cos x \, dx$

12. $\displaystyle\int_0^1 \tan^{-1} x \, dx$ (*Hint*: Use the method of integration by parts.)

13. Determine the change in entropy (*ds*) of 12 kg of iron when its temperature decreases from $T_1 = 800$ K to $T_2 = 300$ K. Use the formula $ds = \displaystyle\int_{T_1}^{T_2} mc \frac{dT}{T}$, where *m* is the mass of iron and $c = 450$ J/kg $\cdot$ K.

14. The wavefunction describring a particle in an infinte potential well of width a is given by $\Psi = A \sin \dfrac{n\pi x}{a}$. The wavefuntcion satisfies the normalization condition $\displaystyle\int_0^a |\Psi|^2 \, dx = 1$. Show that the constant $A = \sqrt{\dfrac{2}{a}}$. [*Hint:* $\displaystyle\int \sin^2 x \, dx = \dfrac{1}{2}x - \dfrac{1}{4}\sin 2x$.]

15. The wavefunction of the ground state of a hydrogen atom is given by the expression $\Psi_{100} = \dfrac{1}{\sqrt{\pi r_0^3}} e^{-r/r_0}$ and r_0 is a constant called the Bohr radius. Show that $\displaystyle\int_0^\infty |\Psi_{100}|^2 \, 4\pi r^2 \, dr = 1$. [*Hint:* $\displaystyle\int x^2 \, e^{-x} dx = (-\dfrac{1}{2}x^2 - x - 1) \, e^{-x}$, which can be verified by integrating by parts.]

16. The average value of the distance r of an electron in the ground state of a hydrogen atom is given by the expression $\langle r \rangle = \displaystyle\int_0^\infty r |\Psi_{100}|^2 \, 4\pi r^2 \, dr$ where $\Psi_{100} = \dfrac{1}{\sqrt{\pi r_0^3}} e^{-r/r_0}$ and r_0 is a constant called the Bohr radius. Show that $\langle r \rangle = \dfrac{3}{2} r_0$. [*Hint:* $\displaystyle\int_0^\infty x^3 e^{-x} \, dx = 3!$]

8.8   Geometric Interpretation of a Definite Integral

The definite integral of the function $y = \phi(x)$ between the limits a and b, that is, $\int_a^b \phi(x)\, dx$, geometrically represents the area of the region enclosed by the curve $y = \phi(x)$, the x-axis, and the ordinates at $x = a$ and b. This was described in Section 8.6.

Example 8.8.1 (from thermal physics)

In an isobaric process, the pressure P of a gaseous system remains constant. As the gas expands, it does work and the work done when its volume increases by an infinitesimal amount dV is given by $dW = P\, dV$. Determine the total work done in an isobaric process when the volume of a gas increases from an initial volume V_1 to a final volume V_2. Show that the total work done is given by the area between the isobaric line and the volume axis in a P-V diagram, where volume of the gas is plotted along x-axis and its pressure is plotted along y-axis.

Since the work in changing the volume by an infinitesimal amount dV is given by dW, the total work done in increasing the volume from V_1 to V_2 is given by the definite integral

$$W = \int_{V_1}^{V_2} dW = \int_{V_1}^{V_2} P\, dV.$$

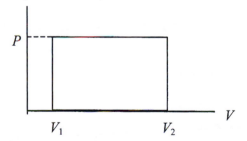

Since P is constant in an isobaric process, it can be taken out from the integration. Thus,

$$W = P \int_{V_1}^{V_2} dV = P(V_2 - V_1).$$

The isobaric process is represented by a horizontal line in a PV diagram since the y-coordinate (that is, pressure P) remains the same whereas the volume changes from V_1 to V_2. Clearly, the product represents the area of the shaded rectangle whose height is P and width $(V_2 - V_1)$. This is what is expected according to the geometric interpretation of the definite integral $\int_{V_1}^{V_2} P\, dV.$

Example 8.8.2 (from thermal physics)

In an isothermal process for an ideal gas $PV = nRT$, where n, R and T are constants. Determine the work done during an isothermal process when the volume changes from V_1 to V_2.

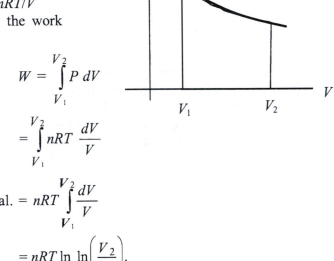

$PV = nRT$ or $P = nRT/V$

As in the previous problem, the work done

$$W = \int_{V_1}^{V_2} P \, dV$$

Substitute $P = nRT/V$.

$$= \int_{V_1}^{V_2} nRT \, \frac{dV}{V}$$

Take nRT out from the integral.

$$= nRT \int_{V_1}^{V_2} \frac{dV}{V}$$

$$= nRT \ln \ln \left(\frac{V_2}{V_1} \right).$$

This work is represented by the shaded area as shown.

Exercise

17. In an adiabatic process for an ideal gas the pressure P and volume V are related by the expression $PV^\gamma = k$ (constant), where γ is a constant of the gas. Show that the work done in an adiabatic process to change the pressure, volume, from P_1, V_1 to P_2, V_2, respectively, is given by $W = (P_1V_1 - P_2V_2)/(\gamma - 1)$. (*Hint*: Use $W = \int_{V_1}^{V_2} P \, dV$ and substitute $P = k/V^\gamma$.)

8.9   Properties of Definite Integrals

Following properties of definite integral are useful in physics. Here the properties are listed without any proof, which can be found in any standard text on integral calculus.

1. $$\int_a^b \phi(x)\, dx = \int_a^b \phi(z)\, dz$$

$$\int_a^a \phi(x)\, dx = 0$$

2. $$\int_a^b \phi(x)\, dx = - \int_b^a \phi(x)\, dx$$

3. $$\int_a^b \phi(x)\, dx = \int_a^c \phi(x)\, dx + \int_c^b \phi(x)\, dx, \text{ for } a < c < b$$

4. $$\int_{+a}^{-a} \phi(x)\, dx = \int_0^a \left[\phi(x) + \phi(-x)\right] dx$$

$$= 2 \int_0^a \phi(x)\, dx, \text{ if } \phi(x) = \phi(-x);\ = 0, \text{ if } \phi(x) = -\phi(-x)$$

5. $$\int_0^a \phi(x)\, dx = \int_0^a \phi(a - x)\, dx$$

Approximate formulas for definite integrals are given in Appendix J.

8.10   Multiple Integrals: Integration over an Area and Volume

In many occasions in physics you will take integrals over an element of area dA or an element of volume dV, as opposed to taking integrals over the element of length dx. Integrals of these kinds are called *multiple integrals*. Although there may be a single integral sign $\int$ for a multiple integral, integral over an area is really a double integral expressed explicitly by the symbol $\iint$ and the integral over a volume is a triple integral expressed explicitly by the symbol $\iiint$.

To perform a multiple integral, the elementary area dA or the elementary volume dV needs to be written as the product of the linear differentials. For example, if Cartesian coordinates are used, $dA = dx\, dy$ (assuming the area dA is in the x-y plane) and $dV = dx\, dy\, dz$. If the function to be integrated is expressed in other coordinates, express the area dA and dV in terms of the linear variables of the same coordinate system. Table 8.1 shows the area dA and dV in different coordinate systems. After you express the area dA or the volume dV as a product of the linear differentials, integrate each differential starting from the innermost differential as if it is a single integral, assuming the other variable or variables to be constant. Then integrate over the other variables one by one following the same strategy and always assuming the other variable(s) constant.

Table 8.1

Coordinate System	Area (dA)	Volume (dV)
Cartesian	$dx\, dy$ $dy\, dz$ $dz\, dx$	$dx\, dy\, dz$
Polar	$r\, dr\, d\theta$	
Cylindrical	$r\, dr\, d\theta$ $dr\, dz$ $r\, d\theta\, dz$	$r\, dr\, d\theta\, dz$
Spherical	$r \sin\phi\, dr\, d\theta$ $r\, dr\, d\phi$ $r^2 \sin\phi\, d\theta\, d\phi$	$r^2 \sin\theta\, dr\, d\theta\, d\phi$

Example 8.10.1

Determine the following multiple integrals:

(a) $\iint (x^2y + 2x + y^3)\, dx\, dy$

(b) $\int_{r=0}^{R} \int_{\theta=0}^{\pi} \int_{\varphi=0}^{2\pi} r^2 \sin\theta\, dr\, d\theta\, d\varphi$

(a) Here dx is the innermost integral, so integrate first over x. To integrate over x, assume y to be a constant. Thus,

$$\iint (x^2y + 2x + y^3)\, dx\, dy$$

$$= \int \left(y \int x^2 dx + 2 \int x\, dx + y^3 \int dx \right) dy$$

$$= \int \left(\frac{1}{3}x^3 y + 2\cdot\frac{1}{2}x^2 + y^3 x + c_1 \right) dy.$$

Next, integrate over y assuming x to be a constant. Thus, the result of the integration is

$$\frac{1}{3}x^3 \cdot \frac{1}{2}y^2 + 2\cdot\frac{1}{2}x^2y + \frac{1}{4}y^4 x + c_1 y + c_2$$

$$= \frac{1}{6}x^3 y^2 + x^2 y + \frac{1}{4}xy^4 + c_1 y + c_2.$$

Note that the result contains two constant c_1 and c_2 because integration was done twice (first over x and then over y) and each integration has given a constant.

(b) $\int_{r=0}^{R} \int_{\theta=0}^{\pi} \int_{\varphi=0}^{2\pi} r^2 \sin\theta\, dr\, d\theta\, d\varphi = \left[\int_{r=0}^{R} r^2 dr \right] \left[\int_{\theta=0}^{\pi} \sin\theta\, d\theta \right] \left[\int_{\varphi=0}^{2\pi} d\varphi \right]$

$$= \frac{r^3}{3}\Big|_0^R \cdot \left[-\cos\theta \right]_0^\pi \cdot \varphi\Big|_0^{2\pi} = \frac{1}{3}R^3 \left[-\cos\pi - 1 \right] 2\pi = \frac{1}{3}R^3 \cdot 2 \cdot 2\pi = \frac{4}{3}\pi R^3$$

Note that in this example the volume element dV [$= r^2 \sin\theta\, dr\, d\theta\, d\varphi$] in spherical coordinate system has been integrated over the all possible values of θ and φ as well as for radius $r = 0$ to R. The three integrations in this example are easily separated; that is, the triple integral is the product of the three individual integrals. When integrated over all possible values of θ and φ (that is, $\theta = 0$ to π and $\varphi = 0$ to 2π and $r = 0$ to R), the result provides the total volume of a sphere of radius R, which is $\frac{4}{3}\pi R^3$, as expected.

Exercises

Determine the following multiple integrals:

18. (a) $\iint (x^2y^2 + 2x + xy^2)\, dx\, dy$ (b) $\int_0^{\pi/2} \int_0^{\sin x} \dfrac{dy\, dx}{\sqrt{1-y^2}}$

19. $\int_{r=0}^{2} \int_{\theta=0}^{\pi} \left((r^2 - 4)r\, dr\, d\theta \right)$

20. $\int_{r=0}^{\infty} \int_{\theta=0}^{\pi} \int_{\phi=0}^{2\pi} \left(\dfrac{1}{\sqrt{\pi r_0^3}} e^{-r/r_0} \right)^2 r^2 \sin\theta\, dr\, d\theta\, d\phi$ (r_0 is a constant)

21. A small volume element dV in cylindrical coordinate system is given by $dV = r\, dr\, d\theta\, dz$. By integrating r from 0 to R, θ from 0 to π and z from 0 to L, show that the volume of a cylinder of radius R and length L is $V = \pi r^3 L$.

Differential Equations

8.11 Differential Equations: Introduction

Many problems of physics are solved by determining how one physical quantity depends on one or more other quantities. Such problems of physics are often solved by means of *differential equations*. Leibniz (1646–1716) and Newton (1642–1727), who invented calculus, initiated the use of differential equations in solving problems.

A differential equation is an equation that contains one or more derivatives, which can be ordinary derivatives, say *dy*/*dx*, or partial derivatives, say $\partial y / \partial x$. For example, the time (*t*) rate of change of the amount *N* of a radioactive substance is proportional to the amount *N*. This radioactive decay can be expressed by the differential equation

$$\frac{dN}{dt} = -\lambda N,$$

where λ is a constant. Many physical laws can be expressed conveniently in terms of differential equations.

An *ordinary differential equation* contains ordinary derivatives of one or more functions with respect to a single independent variable. The preceding differential equation is obviously an ordinary differential equation since it involves the ordinary derivative of *N* with respect to *t*. A *partial differential equation* is one that involves partial derivatives of one or more dependent variables with respect to one or more of the independent variables. For example, the Laplace equation of physics,

$$\nabla^2 f = \frac{\partial^2 f}{\partial x^2} + \frac{\partial^2 f}{\partial y^2} + \frac{\partial^2 f}{\partial x^2} = 0,$$

is a partial differential equation of second order. Another common partial equation in physics is the wave equation,

$$\frac{\partial^2 y}{\partial x^2} = \frac{1}{v^2} \frac{\partial^2 y}{\partial t^2}.$$

The differential equations that you will encounter in physics usually contain first- or second-order derivatives. The *order* of a differential equation is the *order of the highest derivative* in the equation. Between the preceding examples, the differential equation representing radioactive decay is of the first order (since it does not involve any higher derivative beyond first order), while the Laplace equation is of the second order (since it does not involve any derivative beyond the second-order derivative).

A differential equation is said to be *linear* if it is of the *first degree* in the dependent variable and the derivatives. For example, the equation

$$4\frac{d^2 y}{d x^2} + 2\frac{dy}{dx} + 6y = \sin x$$

is linear and of the second order.

Unlike other equations the solution of a differential equation is not a set of numbers that satisfy the equation $y = f(x)$. Instead, the solution is a set of functions $y = f(x)$ that satisfy the differential equation. It can be shown that for any linear differential equation, the *sum of the two solutions is also a solution*. Thus, if y_1 and y_2 are two solutions of a linear differential equation, then $Ay_1 + By_2$ is also a solution, where A and B are arbitrary constants. This property is known as the *superposition principle*.

8.12   Solving Differential Equations

Solving a differential equation, in principle, involves integration. However, the integration can be tricky and difficult depending on the equation. Each kind of differential equation requires a different strategy for a solution. It is not the purpose of this book to discuss all different kinds of differential equations or all the different tricks used for a solution; you should read a text on differential equation for such details. However, I want to focus on solution of the differential equations that are common and that you will encounter in introductory or intermediate physics.

First-Order Ordinary Differential Equations: Separation of Variables

It often happens that an ordinary differential equation of first order appears in one of the forms (or in forms that can be very easily organized in one of these forms):

$$\frac{dy}{dx} = f(x, y),$$

$$M(x, y) + N(x, y)\frac{dy}{dx} = 0.$$

In some cases, the variables x and y are separable and it is possible to divide the equation by a function of x or y or both to make it in the form

$$g(y)\frac{dy}{dx} = h(x),$$

or $g(y)\,dy = h(x)\,dx.$

The variables x and y in the preceding two equations are clearly separated. Passing from the original form of the equation to the preceding form where the variables are separated is known as *separating the variables*.

On integration, we get

$$\int g(y)\, dy = \int h(x)\, dx + c \text{ (integration constant)}.$$

Example 8.12.1

Solve the equation $2x\dfrac{dy}{dx} + y = 0.$

The equation can be written as

$$2x\, dy + y\, dx = 0.$$

To separate the variables x and y, divide both sides by xy.
$$\frac{2x\, dy}{xy} + \frac{y\, dx}{xy} = 0$$

$$\frac{2dy}{y} + \frac{dx}{x} = 0$$

Transfer x to the right hand side.
$$\frac{2dy}{y} = -\frac{dx}{x}$$

Integrate each term on both sides separately.
$$\int \frac{2dy}{y} = -\int \frac{dx}{x}$$

$$2 \ln y = -\ln x + \ln c$$

(integration constant is $\ln c$)

$$\ln(y^2) + \ln x = \ln c$$

$$\ln(xy^2) = \ln c$$

$$xy^2 = c$$

(To verify that the answer makes sense, take the differential of the answer and find out whether you get the original differential equation.
Clearly, $d\,(xy^2) = d\,(c)$

or $\quad x\, d(y^2) + y^2\, dx = 0 \qquad$ or $\qquad x\,(2y\, dy) + y^2\, dx = 0$ or $2xy\, dy + y^2\, dx = 0$

or $\quad$ dividing by $y\, dx$, $2xy\dfrac{dy}{ydx} + y^2\dfrac{dx}{ydx} = 0$ or $\qquad 2x\dfrac{dy}{dx} + y = 0$, as expected.)

Example 8.12.2 (from nuclear physics)

One important first-order ordinary differential equation that appears in physics has the form $dy/dx = ky$, where k is a constant. This equation can be read as *the rate of change of y with respect to x is proportional to y.* For example, as mentioned earlier, the time (t) rate of change of the amount N of a radioactive substance is proportional to the amount N. Thus, radioactive decay follows the differential equation.

$$\frac{dN}{dt} = -\lambda N, \quad \lambda \text{ is a positive constant}$$

We can solve this equation by the method of separation of variables. Here the variables are N and t. Multiply both sides by dt,

$$dN = -\lambda \, N \, dt$$

To separate the variables, divide both sides by N.

$$\frac{dN}{N} = -\lambda \, dt$$

Integrate both sides, knowing that at $t = 0$, $N = N_0$

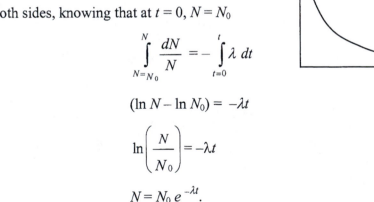

$$\int_{N=N_0}^{N} \frac{dN}{N} = -\int_{t=0}^{t} \lambda \, dt$$

$$(\ln N - \ln N_0) = -\lambda t$$

or

$$\ln\left(\frac{N}{N_0}\right) = -\lambda t$$

or

$$N = N_0 \, e^{-\lambda t}.$$

The accompanying plot shows how N decreases with t from its initial value N_0.

Example 8.12.3 (from electricity)

When the switch of an electrical circuit containing an inductance L and resistnce R and a dc battery of emf V_0 is closed, the growth of current in the circuit follows the following linear differential equation: $L\dfrac{dI}{dt} + RI = V_0$. Solve the equation to determine the current I at a time t after the switch is closed.

Here the variables are I and t. The equation can be solved by using the separation of variables.

$$L\frac{dI}{dt} + RI = V_0.$$

or $$L\frac{dI}{dt} = V_0 - RI$$

Separate the variables. $$\frac{dI}{V_0 - RI} = \frac{dt}{L}$$

or $$\frac{dI}{I - \dfrac{V_0}{R}} = -\frac{R}{L}dt$$

Integrate both sides, remembering that at $t = 0$, $I = 0$.

$$\int_{I=0}^{I} \frac{dI}{I - \dfrac{V_0}{R}} = -\frac{R}{L}\int_{t=0}^{t} dt$$

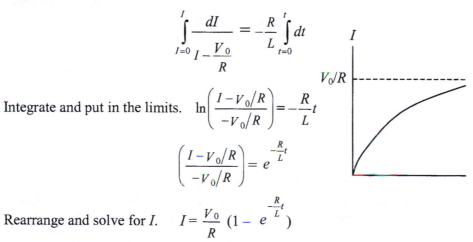

Integrate and put in the limits. $$\ln\left(\frac{I - V_0/R}{-V_0/R}\right) = -\frac{R}{L}t$$

$$\left(\frac{I - V_0/R}{-V_0/R}\right) = e^{-\frac{R}{L}t}$$

Rearrange and solve for I. $$I = \frac{V_0}{R}\left(1 - e^{-\frac{R}{L}t}\right)$$

The graph on the right shows how current I changes with time t.

Example 8.12.4 (from mechanics)

The force F of air resistance on a rapidly falling object has the form $F = -bv$ (where v is velocity and b is a constant) so that Newton's second law for the motion of the object is $m\, dv/dt = mg - bv$. Solve this equation to determine the velocity v as a function of time t and show that the object reaches a *terminal velocity* of magnitude mg/b.

The equation is $$m\frac{dv}{dt} = mg - bv$$

or $$\frac{dv}{dt} = g - \frac{b}{m}v.$$

Separate the variables v and t.
$$\frac{dv}{g - \dfrac{b}{m}v} = dt$$

or
$$\frac{dv}{v - \dfrac{mg}{b}} = -\frac{b}{m}dt$$

Integrate, remembering that at $t = 0$, $v = 0$.

$$\int_{v=0}^{v} \frac{dv}{v - \dfrac{mg}{b}} = -\frac{b}{m}\int_{t=0}^{t} dt$$

Integrate and put in the limits.
$$\ln\left(\frac{v - mg/b}{-mg/b}\right) = -\frac{b}{m}t$$

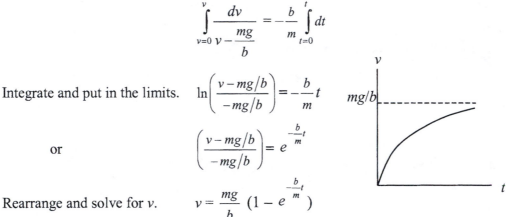

or
$$\left(\frac{v - mg/b}{-mg/b}\right) = e^{-\frac{b}{m}t}$$

Rearrange and solve for v.
$$v = \frac{mg}{b}\left(1 - e^{-\frac{b}{m}t}\right)$$

The preceding equation and the graph at the right above show how the velocity v changes with time t. As t approaches infinity, the term $e^{-bt/m}$ becomes zero. Thus, the velocity approaches a constant value mg/b. This velocity is called the terminal velocity.

Exercises

22. According to Newton's law of cooling, the time (t) rate of change of temperature T of a hot object is proportional to the temperature difference ($T - T_s$), where T_s is the temperature of it surroundings. That is, $dT/dt = -k(T - T_s)$, where k is a positive constant. Show that the solution of the equation is $T = T_s + A\,e^{-kt}$.

23. The change of light intensity I with thickness x of a transparent medium is expressed by the differential equation $dI/dx = -kI$. Show that the light intensity changes with thickness as $I = I_0\,e^{-kx}$, where I_0 is the original intensity.

24. The decay of current I in an electric circuit containing an inductance L and a resistance R follows a differential equation $L\,dI/dt + RI = 0$. Show that the current at time t is given by $I = I_0 e^{-tR/L}$, where I_0 is the initial value of the current.

25. Newton's law for the motion of an object of an object of mass m in presence of a drag force $-bv^{1/2}$ is given by $mdv/dt = -bv^{1/2}$. Solve this equation to show that $v = v_0 - (bv_0^{1/2}m)t + (b^2/4m^2)t^2$, where v_o is the initial velocity of the object.

8.13   Second-Order Ordinary Differential Equation

Motion with a Constant Acceleration

The simplest differential equation of second order that arises in physics is $\dfrac{d^2 y}{d x^2} = c$, where c is a constant. As shown below, this differential equation represents a motion with a constant acceleration. Acceleration of an object gives the time rate of change of its velocity.

For example, consider the motion of an object of mass m falling freely near the surface of the earth. This motion can be expressed by Newton's second law of motion, $F = ma$, which provides a relationship between the force F and the constant acceleration a of the particle. This equation can be expressed by the differential equation

$$m\frac{d^2 y}{d t^2} = mg$$

or

$$\frac{d^2 y}{d t^2} = g,$$

where y is the vertical position of the object at time t and g is the constant acceleration due to gravity. Since velocity is the time rate of change of velocity, $v = dy/dt$ and $a = dv/dt = d(dy/dt)/dt = d^2y/dt^2$. Thus,

$$\frac{dv}{dt} = g$$

Multiply by dt.

$$dv = g\, dt$$

Integrate both sides.

$$\int dv = \int g dt$$

$$v = gt + A \text{ (constant of integration)}.$$

The constant A can be determined from the initial condition of the problem. If at $t = 0$ the velocity, $v = v_0$ (initial velocity), then from the preceding equation, $v_0 = g \cdot 0 + A = A$. Therefore, from the preceding equation,

$$v = v_0 + gt.$$

To determine how the position y changes with time for a freely falling object, rewrite the preceding equation as

$$\frac{dy}{dt} = v_0 + gt$$

Multiply by dt.

$$dy = v_0\, dt + gt\, dt$$

Integrate both sides.

$$y = \int v_0\, dt + \int gt\, dt$$

$$y = v_0 t + \frac{1}{2} gt^2 + A \text{ (constant of integration)}$$

If at $t = 0$, $y = y_0$ (initial height of the object), then from the preceding equation,

$$y_0 = v_0 \cdot 0 + \frac{1}{2} g \cdot 0 + A$$

$$y_0 = A.$$

Therefore,

$$y = y_0 + v_0 t + \frac{1}{2} gt^2.$$

This is the familiar kinematic equation in physics showing how the position y changes with time t for an object moving under the constant acceleration g.

Simple Harmonic Motion: Motion of a Mass Attached to a Spring

The restoring force acting on a mass m for the motion of the mass attached to a spring, is given by Hooke's law $F = -ky$ or, in differential form, $m\frac{d^2 y}{dt^2} = -ky$. Here y represents the amount of stretch or compression of the spring, t represents time, and k is a constant of the spring called spring constant. From the preceding equation,

$$\frac{d^2 y}{dt^2} = -\frac{k}{m} y = -\omega^2 y.$$

This is an important second-order linear differential equation in physics. An equation of the form $\dfrac{d^2 y}{d t^2} = -\omega^2 y$, in general, represents a simple harmonic motion such as the motion of a mass attached to a spring or the motion of a pendulum for small displacements.

This equation has two solutions (as can be verified easily by direct substitution of the derivatives of the solutions), which are $\cos \omega t$ and $\sin \omega t$. (You can verify by direct substitution of $y = \cos \omega t$ or $y = \sin \omega t$ in the preceding equation that these are two solutions of the equation.) Thus, $y = A \cos \omega t + B \sin \omega t$ is also a solution (based on the superposition principle). To verify that the preceding expression for y is a solution, first take the first and second derivatives of y with respect to t.

First derivative, $\dfrac{dy}{dt}$ or y': $A\omega (-\sin \omega t) + B\omega \cos \omega t$

Second derivative, $\dfrac{d^2 y}{d t^2}$, or, y'' $A\omega^2(-\cos \omega t) + B\omega^2(-\sin \omega t)$

Therefore, $\dfrac{d^2 y}{d t^2} = -\omega^2(A\cos \omega t + B\sin \omega t)$

$= -\omega^2 y$, as expected.

Example 8.13.1

Show that the solution $y = A \cos \omega t + B \sin \omega t$ of the preceding differential equation can also be written in the form $y = C \cos (\omega t + \phi)$, where C and ϕ are constants.

Using the identity $\cos (A + B) = \cos A \cos B - \sin A \sin B$, we can rewrite the expression $y = C \cos (\omega t + \phi)$ as

$$y = C \cos (\omega t + \phi)$$

$$= C(\cos \omega t \cos \phi - \sin \omega t \sin \phi)$$

$$= C \cos \omega t \cos \phi - C \sin \omega t \sin \phi$$

$$= (C \cos \phi)\cos \omega t + (- C \sin \phi)\sin \omega t$$

If we let $A = C \cos \phi$ and $B = - C \sin \phi$, then

$$y = A \cos \omega t + B \sin \omega t.$$

Exercises

26. Show by direct substitution that $x = A \cos(\sqrt{\frac{g}{l}}\, t + \phi)$ is a solution of the differential equation $\frac{d^2 x}{dt^2} = -\frac{g}{l} x$. This differential equation represents the motion of a pendulum of length l for small angular displacements.

27. Show by direct substitution that $y = A \sin(2\pi/\lambda)(x \pm vt)$ is a solution of the wave equation, $\frac{\partial^2 y}{\partial x^2} = \frac{1}{v^2}\frac{\partial^2 y}{\partial t^2}$. Here the constants A, λ, and v represent the maximum height or the amplitude of the wave, wavelength of the wave, and velocity of the wave, rspectively.

Damped Harmonic Motion

Damped harmonic motion that occurs in nature can be approximated by restoring force $F = -kx$ (k is called the spring constant) and a damping force $-bv$ where b is the damping constant and $v = dx/dt$, the velocity of the object. The object thus follows the differential equation

$$F = ma = m\frac{d^2 x}{dt^2} = -kx - b\frac{dx}{dt}$$

or $$m\frac{d^2 x}{dt^2} + b\frac{dx}{dt} + kx = 0.$$

Obviously, when the damping constant b is negligible, the equation becomes

$$m\frac{d^2 x}{dt^2} + kx = 0,$$

whose solution (that gives how the displacement x changes with time t) is given by
$$x = A \cos(\omega t + \phi),$$
where $\omega^2 = k/m$ and A is a constant, as discussed earlier.

Clearly, when the cosine function on the right of the preceding equation is unity, $x = A$. Thus, A represents the maximum value of x, called the *amplitude* of motion.
If for the system the displacement $x = A$ when time $t = 0$, we find from the preceding expression $\phi = 0$. Thus, $x = A \cos \omega t$ is a solution of the undamped harmonic motion in this situation.

When the damping constant b is small but not negligible, it is expected that the solution is still a cosine function but times a factor that decreases in time t. A simple function that does this is the exponential $e^{-\alpha t}$, and the solution that satisfies the differential equation of the damped harmonic motion is

$$x = A\, e^{-\alpha t}\cos \omega' t,$$

where A, $\alpha\, (= b/2m)$ and $\omega' = \sqrt{k/m - b^2/4m^2}$ are constants and the displacement $x = A$ at $t = 0$.

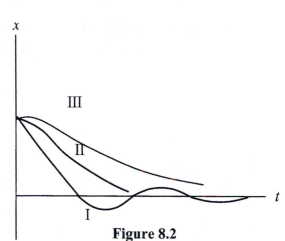

Figure 8.2

- Case I: If the damping is such that $b^2 < 4km$, ω' is real and the system oscillates with the frequency ω' because of the cosine function in the preceding equation and the amplitude of the oscillation decays exponentially. In other words, the system makes several swings before it comes to rest. The system is said to *be underdamped* in this situation. This situation is shown by curve I in Figure 8.2.

- Case II: If $b^2 = 4km$, the system does not oscillate but reaches the equilibrium in the shortest time. The system is said to *critically damped* in this situation. This situation is shown by curve II in Figure 8.2.

- Case III: If the damping is such that $b^2 > 4km$, then ω' is imaginary. That is, in this situation the system does not oscillate but returns to the equilibrium position. If the damping is too high it takes a long time for the system to reach equilibrium. The system is said to be *overdamped* in this situation. This situation is shown by curve III in Figure 8.2.

Exercise

28. Show by direct substitution that $x = A\, e^{-\alpha t}\cos \omega' t$ is a solution of the differential equation $m\, d^2x/dt^2 + b\, dx/dt + kx = 0$, where $\alpha = b/2m$ and $\omega' = \sqrt{k/m - b^2/4m^2}$.

Forced Oscillation: Resonance

A periodic external force $F_{ext} = F \cos \omega t$ is applied to the motion of a damped

harmonic oscillator that otherwise obeys the differential equation $F = ma = m\dfrac{d^2x}{dt^2} =$

$-kx - b\dfrac{dx}{dt}$. Thus, the system obeys the differential equation for the object is

$$F = ma = m\frac{d^2x}{dt^2} = -kx - b\frac{dx}{dt} + F \cos \omega t.$$

The equation can be rearranged as

$$m\frac{d^2x}{dt^2} + b\frac{dx}{dt} + kx = F \cos \omega t.$$

The right-hand side of this equation does not involve x or its derivative. It can be
shown that the solution of the preceding equation is

$$x = A \sin (\omega t + \phi),$$

where A and ϕ are constants. Thus, the system oscillates sinusoidally with amplitude
A. The constant A and ϕ are given by

$$A = \frac{F}{m\sqrt{\left(\omega^2 - \omega_0^2\right)^2 + b^2\omega^2/m}}, \text{ where } \omega_0 = \sqrt{k/m}, \text{ called the } natural\ frequency \text{ of}$$

the system and

$$\phi = \tan^{-1}\frac{\omega^2 - \omega_0^2}{b\omega/m}.$$

Clearly, from the preceding expression of A it is seen that the amplitude strongly depends on the difference between the frequency ω of the applied force and the natural frequency ω_0 of the system. A plot of the amplitude A as a function of the applied frequency ω for the case of light damping (b is small) the amplitude A becomes very large near the natural frequency $\omega = \omega_0$. When damping constant $b = 0$, the amplitude of oscillation is a maximum at $\omega = \omega_0$, as shown in Figure 8.3. This situation is called *resonance*. The phenomenon of resonance is important in physics.

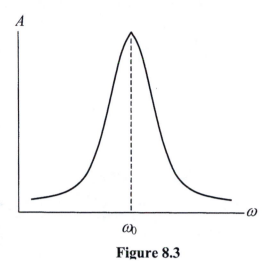

Figure 8.3

The narrowness and height of a resonant peak is specified by a parameter called *quality factor*, often also called the Q value of a resonance. The Q value of a resonance is given by $Q = m\omega_0/b$.

Exercise

29. Show by direct substitution that $x = A \sin(\omega t + \phi)$ is a solution of the differential equation $m \ d^2x/dt^2 + b \ dx/dt + kx = F \cos \omega t$, where $A = F\bigg/ m\sqrt{\left(\omega^2 - \omega_0^2\right)^2 + b^2\omega^2/m}$ and $\phi = \tan^{-1}\left(\omega^2 - \omega_0^2\big/b\omega/m\right)$ with $\omega_0 = \sqrt{k/m}$.

Answers to Exercises in Chapter 8

1. $\dfrac{1}{2}(x - \dfrac{1}{2}\sin 2x)$

2. $\dfrac{1}{2}\left[\dfrac{\sin(m-n)x}{(m-n)} - \dfrac{\sin(m+n)x}{(m+n)}\right]$

3. $\dfrac{1}{2}(\log x)^2 + c$

4. $\tan^{-1}(\sin x) + c$

5. $\dfrac{1}{2}\sin^{-1}(2x/3) + c$

6. $e^x + 1 - \log(e^x + 1) + c$

7. $x \sin x + \cos x + c$

8. $x(\log x - 1) + c$

11. $\dfrac{1}{6}$

12. $\dfrac{\pi}{4} - \dfrac{1}{2}\log 2$

13. 5296 J/K

18. (a) $x^3 y^3/9 + x^2 y^3/6 + x^2 y + c_1 y + c_2$ (b) $\dfrac{\pi^2}{8}$

19. -4π

20. 1

Chapter 9
Vector Calculus

Introduction

Physical quantities that we encounter in physics and applied mathematics can be represented as mathematical functions of several variables. Vector calculus is needed to study the variations or changes of the vector physical quantities.

Vector calculus plays an important role in electricity and magnetism. Laws of electricity and magnetism are formulated in terms of electric and magnetic fields. Electric and magnetic fields are vector quantities. Vector calculus is not the same as calculus for scalars. Like other physical laws, the laws of electricity and magnetism are expressed conveniently in differential equations involving vector derivatives, divergence, and curl of electric and magnetic fields. Basic concepts of vector calculus, including the rules for vector derivatives, gradient, divergence, and curl, and basic theorems of vector calculus are described.

9.1   Vector Derivatives

Derivatives with vectors can be of different forms, depending on whether the function to be derived is a scalar function or a vector function.

The Gradient of a Scalar

Scalar quantities such as temperature or pressure in a room may vary from place to place in the room. Suppose we have a scalar quantity f (such as the temperature of a room) that is a function of position; that is $f = f(x, y, z)$. At some other point that is displaced by a displacement vector $d\mathbf{r}$, the value of the scalar will have changed to $f + df$ (Figure 9.1) where

$$df = \frac{\partial f}{\partial x}\, dx + \frac{\partial f}{\partial y}\, dy + \frac{\partial f}{\partial z}\, dz$$

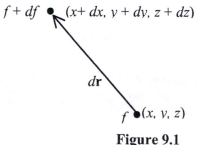

Figure 9.1

Since $d\mathbf{r} = dx\,\mathbf{i} + dy\,\mathbf{j} + dz\,\mathbf{k}$, the preceding expression can be written as

$$df = \left(\frac{\partial f}{\partial x}\mathbf{i} + \frac{\partial f}{\partial y}\mathbf{j} + \frac{\partial f}{\partial z}\mathbf{k}\right) \cdot (dx\,\mathbf{i} + dy\,\mathbf{j} + dz\,\mathbf{k})$$

$$= \nabla f \cdot d\mathbf{r},$$

where $\nabla f = \dfrac{\partial f}{\partial x}\mathbf{i} + \dfrac{\partial f}{\partial y}\mathbf{j} + \dfrac{\partial f}{\partial z}\mathbf{k}$, called the *gradient* of the scalar f. Gradient is a vector quantity with three components. Like any vector, gradient has a magnitude and direction. From the preceding equation, $df = |\nabla f||d\mathbf{r}|\cos\theta$, where θ is the angle between ∇f and $d\mathbf{r}$. For a given $|d\mathbf{r}|$, df is greatest when you move in the same direction as ∇f. Thus, *the gradient points in the direction of the maximum increase of the scalar function f.*

$$\nabla f = \left(\frac{\partial f}{\partial x}\mathbf{i} + \frac{\partial f}{\partial y}\mathbf{j} + \frac{\partial f}{\partial z}\mathbf{k} \right)$$

$$= \left(\frac{\partial}{\partial x}\mathbf{i} + \frac{\partial}{\partial y}\mathbf{j} + \frac{\partial}{\partial z}\mathbf{k} \right) f$$

The term in the parentheses is called "del," ∇. ∇ is a *vector* operator that acts on a function f.

Example 9.1.1

Calculate the gradient of a scale function $f(x, y, z) = x^3 - y + 2z$.

In this case, take the partial derivatives.

$$\frac{\partial f}{\partial x} = 3x^2, \qquad \frac{\partial f}{\partial y} = -1, \qquad \text{and } \frac{\partial f}{\partial z} = 2$$

$$\nabla f = \frac{\partial f}{\partial x}\mathbf{i} + \frac{\partial f}{\partial y}\mathbf{j} + \frac{\partial f}{\partial z}\mathbf{k} = 3x^2\mathbf{i} - \mathbf{j} + 2\mathbf{k}$$

Example 9.1.2

Calculate the gradient of $1/r$, where $r = (x^2 + y^2 + z^2)^{1/2}$ is the magntitude of the position vector $\mathbf{r} = x\mathbf{i} + y\mathbf{j} + z\mathbf{k}$.

In this case,
$$\nabla\left(\frac{1}{r}\right) = \left(\frac{\partial}{\partial x}\mathbf{i} + \frac{\partial}{\partial y}\mathbf{j} + \frac{\partial}{\partial z}\mathbf{k} \right)\left(\frac{1}{r}\right)$$

$$= \left(\frac{\partial}{\partial x}\mathbf{i} + \frac{\partial}{\partial y}\mathbf{j} + \frac{\partial}{\partial z}\mathbf{k} \right)(x^2 + y^2 + z^2)^{-1/2}.$$

Now, $\dfrac{\partial}{\partial x}(x^2 + y^2 + z^2)^{-1/2} = -\dfrac{1}{2}(x^2 + y^2 + z^2)^{-3/2}(2x) = -x(x^2 + y^2 + z^2)^{-3/2}$, and so on.

Thus,
$$\nabla\left(\frac{1}{r}\right) = -(x\mathbf{i} + y\mathbf{j} + z\mathbf{k})\,(x^2 + y^2 + z^2)^{-3/2}$$

$$= -\frac{\mathbf{r}}{r^3}.$$

Example 9.1.3 (from mechanics)

The gravitational force due to a mass M on a mass m is given by $\mathbf{F} = \dfrac{GMm}{r^3}\mathbf{r}$, where r is the distance between the masses. Show that the force can be expressed as the gradient of a scalar.

$$\mathbf{F} = \frac{GMm}{r^3}\mathbf{r} = -(GMm)\nabla\left(\frac{1}{r}\right)\;[\text{since } \nabla\left(\frac{1}{r}\right) = -\frac{\mathbf{r}}{r^3}\,]$$

$$= -\nabla\left(\frac{GMm}{r}\right).$$

The quantity GMm/r is a scalar, say, ϕ. Thus, $\mathbf{F} = -\nabla\phi$, where $\phi = GMm/r$, ϕ is called the gravitational potential energy of the mass m.

As a vector $\mathbf{A}$ can be multiplied in three ways (scalar multiplication, dot product, and cross product), there are three ways the vector operator ∇ can act:

ii. Acts on a scalar function f to provide the *gradient* ∇f

ii. Acts on a vector function $\mathbf{A}$ in the form of a dot product $\nabla \cdot \mathbf{A}$ (called the *divergence*), and

iii. Acts on a vector function $\mathbf{A}$ in the form of a cross product $\nabla \times \mathbf{A}$ (called the *curl*).

The Divergence of a Vector

The divergence of a vector **A** is defined as

$$\nabla \cdot \mathbf{A} = \left(\frac{\partial}{\partial x}\mathbf{i} + \frac{\partial}{\partial y}\mathbf{j} + \frac{\partial}{\partial z}\mathbf{k} \right) \cdot (A_x\mathbf{i} + A_y\mathbf{j} + A_z\mathbf{k})$$

$$= \frac{\partial A_x}{\partial x} + \frac{\partial A_y}{\partial y} + \frac{\partial A_z}{\partial z}.$$

Since the dot product is a scalar, the *divergence* of a vector is a *scalar*. Divergence of a scalar does not have any meaning. The divergence gets its name from association with the motion of a fluid. It can be shown that the divergence of the velocity of a fluid is equal to the volume of the fluid that diverges from a unit volume in unit time.

Note: Geometrically, the divergence of a vector means how much the vector spreads or diverges at the point. If the divergence is *positive* at a point, then there is *source* at the point. If the divergence of a vector is *negative*, there is *a sink* at the poin If the divergence is *zero*, there is neither a *source nor a* sink at the point. Thes situations are shown at the points marked as ● in Figure 9.2.

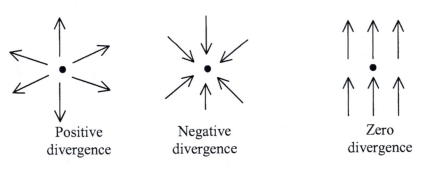

| Positive divergence | Negative divergence | Zero divergence |

Figure 9.2

- If $\nabla \cdot \mathbf{A} = 0$, then the vector field **A** is called a solenoidal vector field. For example, magnetic field is a vector and $\nabla \cdot \mathbf{B} = 0$ always for the magnetic field **B**. Thus, **B** is a solenoidal vector field.

- It can be shown that the *divergence of the curl of a vector* is zero; that is, for a vector $\mathbf{\Phi}$, $\nabla \cdot (\nabla \times \mathbf{\Phi}) = 0$. This means that if $\nabla \cdot \mathbf{A} = 0$ for a vector $\mathbf{A}$, then the vector $\mathbf{A}$ can be obtained from the vector $\mathbf{\Phi}$ so that $\mathbf{A} = \nabla \times \mathbf{\Phi}$. $\mathbf{\Phi}$ is called the vector potential.

Example 9.1.4

Determine the divergence of the position vector $\mathbf{r} = (x\,\mathbf{i} + y\,\mathbf{j} + z\mathbf{k})$.

In this case, $\dfrac{\partial A_x}{\partial x} = \dfrac{\partial x}{\partial x} = 1$, $\dfrac{\partial A_y}{\partial y} = \dfrac{\partial y}{\partial y} = 1$, and $\dfrac{\partial A_y}{\partial z} = \dfrac{\partial z}{\partial z} = 1$.

$$\nabla \cdot \mathbf{r} = \frac{\partial x}{\partial x} + \frac{\partial y}{\partial y} + \frac{\partial z}{\partial z} = 1 + 1 + 1 = 3$$

It is seen from Figure 9.2 that the first vector function in the figure can be expressed as $\mathbf{A} = \mathbf{r}$ since each vector points radially outward from the origin (the source). Thus, $\nabla \cdot \mathbf{A} = \nabla \cdot \mathbf{r} = 3$ (see Example 9.1.4). This vector function thus has a positive divergence, as expected. The second vector function in Figure 9.2 can be expressed as $\mathbf{A} = -\mathbf{r}$ and $\nabla \cdot \mathbf{A}$ is negative. The third function in Figure 9.2 can be expressed as $\mathbf{A} = a\mathbf{k}$, where a is a positive scalar since the vectors are pointing in the z-direction. Thus,

$$\nabla \cdot \mathbf{A} = \frac{\partial(0)}{\partial x} + \frac{\partial(0)}{\partial y} + \frac{\partial(a)}{\partial z} = 0 + 0 + 0 = 0, \text{ as expected.}$$

*Example 9.1.5 (from electrostatics)

The elctrostatic field vector at a point r due to a point electric charge q is given by

$\mathbf{E} = \dfrac{q}{4\pi\varepsilon_0}\dfrac{1}{r^2}\hat{r}$, where ε_0 is a constant and $\hat{r}$ is a unit vector along r.

Show that $\nabla \cdot \mathbf{E} = \mathbf{0}$.

$$\mathbf{E} = \frac{q}{4\pi\varepsilon_0}\frac{1}{r^2}\hat{r} = \frac{q}{4\pi\varepsilon_0}\frac{\mathbf{r}}{r^3}; \text{ then } \nabla \cdot \mathbf{E} = \frac{q}{4\pi\varepsilon_0}\nabla \cdot \left(\frac{\mathbf{r}}{r^3}\right).$$

Use the vector identity $\nabla \cdot (f\mathbf{A}) = \mathbf{A} \cdot (\nabla f) + f(\nabla \cdot \mathbf{A})$ (given in Table 9.1). Here $f = 1/r^3$ and $\mathbf{A} = \mathbf{r}$. Therefore,

$$\nabla \cdot \mathbf{E} = \frac{q}{4\pi \varepsilon_0} \left[\mathbf{r} \cdot \nabla \left(\frac{1}{r^3} \right) + \frac{1}{r^3} \nabla \cdot \mathbf{r} \right].$$

Now, $\nabla \left(\dfrac{1}{r^3} \right) = -3 \dfrac{\mathbf{r}}{r^5}$; this is left as an exercise in this chapter and can be shown

following Example 9.1.2. Also, $\nabla \cdot \mathbf{r} = 3$ (Example 9.1.4).

Thus, $\nabla \cdot \mathbf{E} = \dfrac{q}{4\pi \varepsilon_0} \left[\mathbf{r} \cdot \left(-3 \dfrac{\mathbf{r}}{r^5} \right) + \dfrac{1}{r^3} (3) \right] = \dfrac{q}{4\pi \varepsilon_0} \left[-\dfrac{3}{r^3} + \dfrac{3}{r^3} \right] = 0.$

The Curl of a Vector

The curl of a vector $\mathbf{A}$ is defined as

$$\nabla \times \mathbf{A} = \left(\frac{\partial}{\partial x} \mathbf{i} + \frac{\partial}{\partial y} \mathbf{j} + \frac{\partial}{\partial z} \mathbf{k} \right) \times (A_x \mathbf{i} + A_y \mathbf{j} + A_z \mathbf{k})$$

Express in determinant form.
$$= \begin{vmatrix} \mathbf{i} & \mathbf{j} & \mathbf{k} \\ \dfrac{\partial}{\partial x} & \dfrac{\partial}{\partial y} & \dfrac{\partial}{\partial z} \\ A_x & A_y & A_z \end{vmatrix}$$

Expand the determinant.
$$= \left(\frac{\partial A_x}{\partial y} - \frac{\partial A_y}{\partial z} \right) \mathbf{i} + \left(\frac{\partial A_x}{\partial z} - \frac{\partial A_z}{\partial x} \right) \mathbf{j} + \left(\frac{\partial A_y}{\partial x} - \frac{\partial A_x}{\partial y} \right) \mathbf{k}.$$

Since the cross product of two vectors is a vector, the *curl* of a vector is a *vector*. The curl of a scalar is not meaningful. The curl is especially used in the study of rotational motion because it relates the linear velocity of particles of a rigid body to the angular velocity of the body. Curl is alsoo used in electromagnetic theory because it relates a magnetic field to the current that it produces.

Note: Geometrically, the curl of a vector is a measure of how the vector curls around at the point of interest. For example, the velocity of water on the surface of water at a point has a nonzero curl if you spread some sawdust at the point and you find that it starts rotating. The vector function in Figure 9.3 has a nonzero curl, which points in the z-direction according to the right-hand rule. That the curl is nonzero for the vector funtion of Figure 9.3 can be shown mathematically from the preceding definition of the curl, and this is left as an exercise in this chapter.

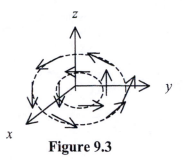

Figure 9.3

- If the curl of a vector **A** is zero everywhere in a vector field, the vector field is called *irritoational*. For example, if the curl of the velocity vector **v** is zero everywhere in a fluid, the fluid motion is said to be irrotational. In such a case, about a given point $\nabla \times \mathbf{v} = 0$ and the particles of the fluid will have no angular velocity about the point.

- Also, it can be shown that *the curl of the gradient* of a scalar is zero. That is, $\nabla \times (\nabla \phi) = 0$, where ϕ is a scalar function. Thus, if $\nabla \times \mathbf{A} = 0$ for a vector **A,** then **A** can be obtained from a scalar function ϕ (called scalar potential) such that $\mathbf{A} = -\nabla \phi$. For example, the gravitational force **F** acting on a particle of mass m or the electrostatic force acting on a particle carrying an electric charge q can be expressed in terms of the corresponding gravitational or electrostatic potential energy ϕ according to the relation $F = -\nabla \phi$ (see Example 9.1.3). Thus for the gravitational force and the electrostatic force $\nabla \times \mathbf{F} = 0$. In general, any force **F** for which $\nabla \times \mathbf{F} = 0$ is called a *conservative* force. Gravitational and electrostatic forces are examples of conservative forces.

> Note: Sprinkle some fine particles that float on the surface of water and observe the motion of the particles on water. If you find that the particles you sprinkled spread out, then you have dropped them on the point having positive divergence. If you find that the particles move toward a common point (called sink) to meet, then you have dropped them on the point having negative divergence. If you find that the particles start to rotate, then you have dropped them on the point having nonzero curl.

Example 9.1.6

Determine the curl of the vector function $\mathbf{A} = 2x^2 \, \mathbf{i} + 3x \, \mathbf{j}$.

In this case,

$$\nabla \times \mathbf{A} = \begin{vmatrix} \mathbf{i} & \mathbf{j} & \mathbf{k} \\ \dfrac{\partial}{\partial x} & \dfrac{\partial}{\partial y} & \dfrac{\partial}{\partial z} \\ A_x & A_y & A_z \end{vmatrix} = \begin{vmatrix} \mathbf{i} & \mathbf{j} & \mathbf{k} \\ \dfrac{\partial}{\partial x} & \dfrac{\partial}{\partial y} & \dfrac{\partial}{\partial z} \\ 2x^2 & 3x & 0 \end{vmatrix}$$

$$= \left(\frac{\partial(0)}{\partial y} - \frac{\partial(3x)}{\partial z} \right) \mathbf{i} + \left(\frac{\partial(2x^2)}{\partial z} - \frac{\partial(0)}{\partial x} \right) \mathbf{j} + \left(\frac{\partial(3x)}{\partial x} - \frac{\partial(2x^2)}{\partial y} \right) \mathbf{k}$$

$$= (0 - 0)\mathbf{i} + (0 - 0)\mathbf{j} + (3 - 0)\mathbf{k}.$$

(Note that the partial derivatives of $2x^2$ and $3x$ are zero with respect to y and z.)

Thus, $\nabla \times \mathbf{A} = 0\mathbf{i} + 0\mathbf{j} + 3\mathbf{k} = 3\mathbf{k}$.

Example 9.1.7 (from mechanics)

Show that for a rigid body in motion, the curl of its liner velocity ($\mathbf{v}$) at any point ($\mathbf{r}$) is equal to twice its angular velocity $\boldsymbol{\omega}$. Use the relation $\mathbf{v} = \boldsymbol{\omega} \times \mathbf{r}$.

In this case,
$$\mathbf{v} = \boldsymbol{\omega} \times \mathbf{r} = \begin{vmatrix} \mathbf{i} & \mathbf{j} & \mathbf{k} \\ \omega_x & \omega_y & \omega_z \\ x & y & z \end{vmatrix}.$$

Expand the determinant.
$$= (\omega_y z - \omega_z y)\mathbf{i} + (\omega_z x - \omega_x z)\mathbf{j} + (\omega_x y - \omega_y x)\mathbf{k}$$

Therefore,
$$\nabla \times \mathbf{v} = \begin{vmatrix} \mathbf{i} & \mathbf{j} & \mathbf{k} \\ \dfrac{\partial}{\partial x} & \dfrac{\partial}{\partial y} & \dfrac{\partial}{\partial z} \\ \omega_y z - \omega_z y & \omega_z x - \omega_x z & \omega_x y - \omega_y x \end{vmatrix}$$

Expand the determinant.
$$= [(\omega_x - (-\omega_x)]\mathbf{i} + [(\omega_y - (-\omega_y)]\mathbf{j} + [(\omega_z - (-\omega_z)]\mathbf{k}$$

$$= 2(\omega_x \mathbf{i} + \omega_y \mathbf{j} + \omega_y \mathbf{k}) = 2\boldsymbol{\omega}$$

Second Derivatives

Gradient, divergence, and curl are all first derivatives. By applying ∇ in different possible ways on gradient, divergence, and curl, we get the different second derivatives, some of which are common in physics and are as follows:

i. The divergence of gradient: $\nabla \cdot (\nabla f)$

$$\nabla \cdot (\nabla f) = \left(\frac{\partial}{\partial x}\mathbf{i} + \frac{\partial}{\partial y}\mathbf{j} + \frac{\partial}{\partial z}\mathbf{k} \right) \cdot \left(\frac{\partial f}{\partial x}\mathbf{i} + \frac{\partial f}{\partial y}\mathbf{j} + \frac{\partial f}{\partial z}\mathbf{k} \right)$$

$$\frac{\partial^2 f}{\partial x^2} + \frac{\partial^2 f}{\partial y^2} + \frac{\partial^2 f}{\partial x^2} = \nabla^2 f, \text{ where}$$

$$\nabla^2 = \frac{\partial^2}{\partial x^2} + \frac{\partial^2}{\partial y^2} + \frac{\partial^2}{\partial x^2} \text{ is called the Laplacian operator.}$$

ii. The curl of a gradient, $\nabla \times (\nabla f)$, which is always zero.

iii. The divergence of a curl, $\nabla \cdot (\nabla \times \mathbf{A})$, which is always zero.

iv. The curl of curl, $\nabla \times (\nabla \times \mathbf{A})$. It can be shown that

$$\nabla \times (\nabla \times \mathbf{A}) = \nabla (\nabla \cdot \mathbf{A}) - \nabla^2 \mathbf{A}.$$

The use of gradient, divergence, curl, and Laplacian are common in physics in Cartesian coordinate systems as well as in spherical and cylindrical systems. Their expressions in spherical and cylindrical systems are given in appendix I.

Important identities with gradient, divergence, and curl are listed in Table 9.1. Rules of vector derivatives look similar to the rules of scalar derivatives and are mentioned in Table 9.2. In these cases, the dots and cross have their special meanings.

Example 9.1.8

Show that $\nabla^2 f = 0$ for $f(x, y, z) = 3x^2 + 2y^2 - 5z^2$.

Take twice partial derivatives with respect to x, $\frac{\partial f}{\partial x} = 6x$, $\frac{\partial^2 f}{\partial x^2} = 6$.

Take twice partial derivatives with respect to y, $\dfrac{\partial f}{\partial y} = 4y$, $\dfrac{\partial^2 f}{\partial y^2} = 4$.

Take twice partial derivatives with respect to z, $\dfrac{\partial f}{\partial z} = -10z$, $\dfrac{\partial^2 f}{\partial z^2} = -10$.

Therefore, $\nabla^2 f = \dfrac{\partial^2 f}{\partial x^2} + \dfrac{\partial^2 f}{\partial y^2} + \dfrac{\partial^2 f}{\partial x^2} = 6 + 4 - 10 = 0.$

Table 9.1 Identities with Grad, Div, and Curl

Summation Rules

$\nabla (f + g) = \nabla f + \nabla g$

$\nabla \cdot (\mathbf{A} + \mathbf{B}) = \nabla \cdot \mathbf{A} + \nabla \cdot \mathbf{B}$

$\nabla \times (\mathbf{A} + \mathbf{B}) = \nabla \times \mathbf{A} + \nabla \times \mathbf{B}$

Second Derivatives Rules

$\nabla \times (\nabla f) = 0$

$\nabla \cdot (\nabla \times \mathbf{A}) = 0$

$\nabla \times (\nabla \times \mathbf{A}) = \nabla (\nabla \cdot \mathbf{A}) - \nabla^2 \mathbf{A}$

Product Rules

$\nabla (fg) = f \nabla g + g \nabla f$

$\nabla (\mathbf{A} \cdot \mathbf{B}) = \mathbf{A} \times (\nabla \times \mathbf{B}) + \mathbf{B} \times (\nabla \times \mathbf{A}) + (\mathbf{A} \cdot \nabla)\mathbf{B} + (\mathbf{B} \cdot \nabla)\mathbf{A}$

$\nabla \cdot (f\mathbf{A}) = \mathbf{A} \cdot (\nabla f) + f(\nabla \cdot \mathbf{A})$

$\nabla \cdot (\mathbf{A} \times \mathbf{B}) = \mathbf{B} \cdot (\nabla \times \mathbf{A}) - \mathbf{A} \cdot (\nabla \times \mathbf{B})$

$\nabla \times (f\mathbf{A}) = f(\nabla \times \mathbf{A}) - \mathbf{A} \times (\nabla f)$

$\nabla \times (\mathbf{A} \times \mathbf{B}) = (\mathbf{B} \cdot \nabla)\mathbf{A} - (\mathbf{A} \cdot \nabla)\mathbf{B} + \mathbf{A} (\nabla \cdot \mathbf{B})\mathbf{A} - \mathbf{B}(\nabla \cdot \mathbf{A})$

Table 9.2 Rules for Vector Identities

$$\frac{d}{ds}(f\mathbf{A}) = \frac{df}{ds}\mathbf{A} + f\frac{d\mathbf{A}}{ds}$$

$$\frac{d}{ds}(\mathbf{A} \cdot \mathbf{B}) = \mathbf{A} \cdot \frac{d\mathbf{B}}{ds} + \mathbf{B} \cdot \frac{d\mathbf{A}}{ds}$$

$$\frac{d}{ds}(\mathbf{A} \times \mathbf{B}) = \mathbf{A} \times \frac{d\mathbf{B}}{ds} + \frac{d\mathbf{A}}{ds} \times \mathbf{B}$$

Exercises

1. Find the gradient of the following scalar functions:
   (a) $f(x, y, z) = x - 2y + 4z$ (b) $f(x, y, z) = xy + yz + zx$

   (c) $f(x, y, z) = r^2 = x^2 + y^2 + z^2$, where **r** is the position vector of a point in space

2. Show that $\nabla\left(\dfrac{1}{r^3}\right) = -3\dfrac{\mathbf{r}}{r^5}$, where **r** is the position vector of a point in space.

3. The elctrostatic field vector at a point r due to a point electric charge q is given by
   $\mathbf{E} = \dfrac{q}{4\pi\varepsilon_0}\dfrac{1}{r^2}\hat{r}$. Show that electric field can be written as $\mathbf{E} = -\nabla\phi$.

4. Find the Laplacian of the of the following scalar functions:

   (a) $f(x, y, z) = x - 2y + 4z$ (b) $f(x, y, z) = xy + yz + zx$

5. Find the divergence of the following vector functions:
   (a) $xy\mathbf{i} + yz\mathbf{j} + zx\mathbf{k}$ (b) $x^2\mathbf{i} + 2xz\mathbf{j} - zx\mathbf{k}$

6. Show that $\nabla \cdot (r^n\mathbf{r}) = 0$ only for $n = 3$, where **r** is the position vector of a point in space. [*Hint*: Use the product rule, $\nabla \cdot (f\mathbf{A}) = \mathbf{A} \cdot (\nabla f) + f(\nabla \cdot \mathbf{A})$.]

7. Find the curl of the vector functions mentioned in Exercise 5.

8. The vector function in Figure 9.3 can be expressed as $\mathbf{A} = -y\mathbf{i} + x\mathbf{j}$. Show that $\nabla \times \mathbf{A}$ is not zero, but it has zero divergence (that is, $\nabla \cdot \mathbf{A} = 0$).

9. Show that $\nabla \times \mathbf{F} = 0$ for the vector $\mathbf{F} = (2xy + z^3)\mathbf{i} + x^2\mathbf{j} + 3xz^2)\mathbf{k}$.

10. The vector $\mathbf{A} = \hat{r}/r^2$ has a zero divergence (as is seen from Example 9.1.5). Show that the vector also has a zero curl.

9.2   Vector Integration

The following two types of vector integrations are used in introductory physics. These are called *line integration* and *surface integration*. The line integral is used for the calculation of magnetic field by using Ampere's law and for the calculation of work. Surface integration is used to determine the electric field using Gauss's law.

Line Integration

The line integral means the integration of an arbitrary vector $\mathbf{A}$ over a line segment, denoted by $\int_{\text{line}} \mathbf{A} \cdot \mathbf{dr}$, or over a closed path denoted by $\oint_{\text{loop}} \mathbf{A} \cdot \mathbf{dr}$ where $\mathbf{dr}$ is an infinitesimally small line vector in the direction of motion along the line or the loop. A finite line segment from point a to point b (showing the vector $\mathbf{A}$ and the infinitesimally small line vector $\mathbf{dr}$), a closed path, and a surface area are shown in Figure 9.4.

$$\mathbf{A} \cdot \mathbf{dr} = (A_x \mathbf{i} + A_y \mathbf{j} + A_z \mathbf{k}) \cdot (dx\, \mathbf{i} + dy\, \mathbf{j} + dz\, \mathbf{k}) = A_x\, dx + A_y\, dy + A_z\, dz.$$

Thus, $$\int_{\text{line}} \mathbf{A} \cdot \mathbf{dr} = \int_{\text{line}} \left(A_x\, dx + A_y\, dy + A_z\, dz \right).$$

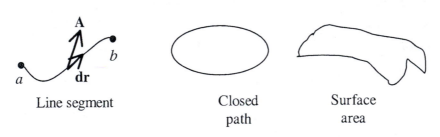

Line segment   Closed   Surface
                       path   area

Figure 9.4

Example 9.2.1

Calculate the line integral of the vector $\mathbf{A} = 2x^2 \mathbf{i} + 3x\, \mathbf{j}$ over the line segment created by a parabola $y = x^2$ from $(0, 0)$ to $(1, 1)$.

In this case, $$\mathbf{A} \cdot \mathbf{dr} = A_x\, dx + A_y\, dy = 2x^2 dx + 3x\, dy.$$

To write dy in terms of dx, use $y = x^2$ so that $dy = 2x\, dx$.

Therefore, $$\mathbf{A} \cdot \mathbf{dr} = 2x^2 dx + 3x\, (2x\, dx) = 8x^2\, dx.$$

Now since the integrand contains x only, integrate it from $x = 0$ to 1.

Therefore,
$$\int_{line} \mathbf{A} \cdot \mathbf{dr} = 8 \int_0^1 x^2 dx = 8 \left. \frac{x^3}{3} \right|_0^1 = \frac{8}{3}.$$

Example 9.2.2 (from mechanics)

The force needed to stretch a nonlinear spring by an amount x from its normal length is given by $F = (kx + ax^3)$, where k and a are constants. Determine the work done to stretch it by an amount X starting from $x = 0$.

As described in Chapter 6, the work done by a variable force is given by

$W = \int_1^2 \mathbf{F} \cdot \mathbf{dr}$. In this case, $\mathbf{F} = (kx + ax^3)\mathbf{i}$ and $\mathbf{dr} = dx\,\mathbf{i}$.

Substitute the values of $\mathbf{F}$ and $\mathbf{dr}$.
$$W = \int_0^X \left(kx + ax^3\right)\mathbf{i} \cdot dx\,\mathbf{i}$$

Separate the integrals.
$$= \int_0^X kx\,dx + + \int_0^X ax^3\,dx$$

$$= k\left.\frac{x^2}{2}\right|_0^X + a\left.\frac{x^4}{4}\right|_0^X = \frac{1}{2}kX^2 + \frac{1}{4}aX^4$$

Example 9.2.3 (from electromagnetism)

When an electric current flows through a wire it generates a magnetic field, which is a vector. Ampere's law of electromagnetism states that the line integral of the magnetic field $\mathbf{B}$ around a closed loop $\oint_{loop} \mathbf{B} \cdot \mathbf{dr} = \mu_0 I$, where I is the current enclosed by the loop and μ_0 is a constant. Use the law to determine the magnetic field B due to a long, straight, cylindrical wire at an outside point distant r from the wire.

Since the wire is long, straight, and cylindrical, because of the symmetry of the situation, the magnetic field depends only on the distance.

Let us choose a closed circular path (shown as dotted line) of radius r concentric with the wire. $\mathbf{B}$ is parallel to the element $\mathbf{dr}$ at any point on the

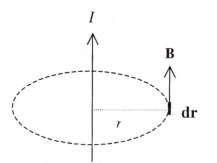

circular path and has constant magnitude. Thus,

$$\oint_{\text{loop}} \mathbf{B} \cdot \mathbf{dr} = \mu_0\, I.$$

B is parallel to **dr**.

$$\oint_{\text{loop}} B\, dr = \mu_0\, I$$

Take out B outside the integral.

$$B \oint_{\text{loop}} dr = \mu_0\, I$$

Length of the closed loop is $2\pi r$.

$$B(2\pi r) = \mu_0\, I$$

Divide by $2\pi r$.

$$B = \frac{\mu_0 I}{2\pi r}$$

Surface Integration

The surface integral means the integration of an arbitrary vector **A** over an open surface, denoted by $\int_{\text{surface}} \mathbf{A} \cdot \mathbf{ds}$ (Figure 9.5), or over a closed surface denoted by $\oint_{\text{surface}} \mathbf{A} \cdot \mathbf{ds}$, where **ds** is an infinitesimally small area vector directed toward the normal to the surface at the point.

$$\mathbf{A} \cdot \mathbf{ds} = (A_x\mathbf{i} + A_y\mathbf{j} + A_z\mathbf{k}) \cdot (ds_x\, \mathbf{i} + ds_y\, \mathbf{j} + ds_z\, \mathbf{k})$$
$$= A_x\, ds_x + A_y\, ds_y + A_z\, ds_z$$

Thus, $\int_{\text{surface}} \mathbf{A} \cdot \mathbf{ds} = \int_{\text{surface}} \left(A_x\, ds_x + A_y\, ds_y + A_z\, ds_z \right).$

Figure 9.5

Surface integration is very useful in electrostatics to calculate electric field of a symmetric charge distribution such as due to a sphere, spherical shell, and a cylinder. An example is given below.

Example 9.2.4 (from electrostatics)

Gauss's law in electrostatics states that the surface integral of the electric field vector, **E**, obeys the relation $\oint_{\text{surface}} \mathbf{E} \cdot \mathbf{ds} = q/\varepsilon_0$, where q is the electric charge enclosed by the surface and ε_0 is a constant. Use this law to determine the electric field at a distance r outside a nonconducting sphere of radius r_0, where an electric charge q is uniformly distributed over the sphere.

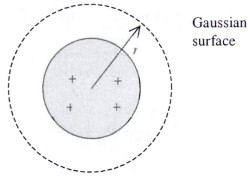

Gaussian surface

Because of the symmetry of a sphere, electric field depends only on the distance and is directed radially outward.

Consider a spherical surface (called Gaussian surface) of radius r (shown as dotted lines in the figure at right) with the center as the center of the sphere. Then

$$\oint_{\text{surface}} \mathbf{E} \cdot \mathbf{ds} = \frac{q}{\varepsilon_0}$$

E and **ds** act in the same direction (radially outward).
$$\oint_{\text{surface}} E\, ds = \frac{q}{\varepsilon_0}$$

Take out E from inside.
$$E \oint_{\text{surface}} ds = \frac{q}{\varepsilon_0}$$

Surface area of a sphere of radius r is $4\pi r^2$.
$$E\,(4\pi r^2) = \frac{q}{\varepsilon_0}$$

Divide by $4\pi r^2$.
$$E = \frac{1}{4\pi \varepsilon_0} \frac{q}{r^2}$$

Note: Gauss's law $\oint_{\text{surface}} \mathbf{E} \cdot \mathbf{ds} = q/\varepsilon_0$ is always true but is useful to determine electric field *only* when there is a *symmetry*. When there is a *spherical symmetry*, make your Gaussian surface a *concentric sphere* (as shown in Example 9.2.4); when there exists a *cylindrical symmetry*, make your Gaussian surface a *coaxial cylinder*; and when there exists a *plane sym*metry, use a Gaussian *pillbox* that straddles the surface.

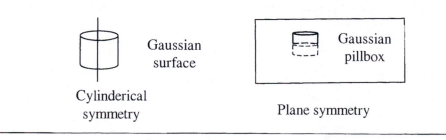

Gaussian surface

Cylinderical symmetry

Gaussian pillbox

Plane symmetry

Exercises

11. Calculate the line integral of the vector $\mathbf{A} = y\mathbf{i} - 3y^2\mathbf{j}$ over the line segment created by a parabola $x = y^2$ from $(0, 0)$ to $(1, 1)$.

12. The total flux of a magnetic field vector $\mathbf{B}$ through a surface is given by $\Phi = \int\limits_{\text{surface}} \mathbf{B} \cdot \mathbf{ds}$. A circular flat coil of wire of area 0.01 m^2 encloses field of 2.0 T that penetrates perpendicularly to the entire surface. Determine the total flux of the magnetic field through the entire area.

13. Apply Gauss's law in electrostatics, $\oint\limits_{\text{surface}} \mathbf{E} \cdot \mathbf{ds} = q/\varepsilon_0$, to show that the electric field E, due to a very long straight wire having a uniform positive charge per unit length λ, at points near but outside the wire is given by $E = (1/2\pi\varepsilon_0)\,\lambda/r$, where r is the distance of the points. (*Hint*: There exists a *cylindrical symmetry*; that is, field E is same at all points on a cylindrical Gaussian surface with the wire as the axis of the cylinder.)

14. Apply Gauss's law in electrostatics, $\oint\limits_{\text{surface}} \mathbf{E} \cdot \mathbf{ds} = q/\varepsilon_0$, to show that the electice field E outside a large uniformly nonconducting surface of area with a surface charge density σ (charge per unit area) is given by $E = \sigma/2\varepsilon_0$. (*Hint*: There exists a *plane symmetry*).

9.3   Fundamental Theorems for Gradient, Divergence, and Curl

The following vector theorems are useful in physics.

The Fundamental Theorem for Gradient

The fundamental theorem for gradient states that the line integral along any selected curve of the gradient is equal to the value of the function at the boundaries. That is,

$$\int_a^b (\nabla f) \cdot d\mathbf{r} = f(b) - f(a).$$

Clearly, the integral is independent of the path from a to b because it depends only on

the endpoints a and b. Also, clearly for a closed path $\oint\limits_{loop} \nabla f \cdot \mathbf{dr} = 0$ since in such a case

$f(b) = f(a)$, as the endpoints are the same.

Example 9.3.1

Work done in moving a particle from
one point to another in general may
depend on the path. For example, the
work done in moving a prticle in a
viscus fluid from a point a to a point b
will be different for different paths
connecting a and b. But for a
conservative force $\mathbf{F} = -\nabla\phi$.

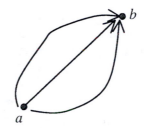

Therefore, the work done W against a conservative force in moving a particle from the

point a to the point b is given by

$$W = -\int_a^b (-\nabla\phi)\cdot \mathbf{dr} = \int_a^b \left(\frac{\partial\phi}{\partial x}\mathbf{i} + \frac{\partial\phi}{\partial y}\mathbf{j} + \frac{\partial\phi}{\partial z}\mathbf{k} \right)\cdot (dx\,\mathbf{i} + dy\,\mathbf{j} + dz\,\mathbf{k})$$

$$= \int_a^b \left(\frac{\partial\phi}{\partial x}dx + \frac{\partial\phi}{\partial y}dy + \frac{\partial\phi}{\partial z}dz \right) = \int_a^b d\phi = \phi(b) - \phi(a).$$

Thus, the work depends only on the values of the potential energy ϕ at the endpoints a
and b, and hence independent of the path.

*Gauss's Divergence Theorem

Gauss's divergence theorem states that the volume integral of the divergence of a
vector over a volume is equal to the surface integral of the vector across the surface
area enclosed by the volume.   That is,

$$\int\limits_{volume} (\nabla\cdot\mathbf{A})\,dV = \oint\limits_{surface} \mathbf{A}\cdot\mathbf{ds},$$

where $dV = dx\,dy\,dz$ in the Cartesian coordinate system.   Note that the volume integral

is really a triple integral.

The surface element **ds** in Cartesian coordinates is $\mathbf{ds}_1 = dy\,dz\,\mathbf{i}$ on the front surface of the cube (Figure 9.6), $\mathbf{ds}_2 = dz\,dx\,\mathbf{j}$ on the right surface and $\mathbf{ds} = dx\,dy\,\mathbf{k}$ for the top surface, whereas for the back face, for example, $\mathbf{ds} = -dy\,dz\,\mathbf{i}$, and so on.

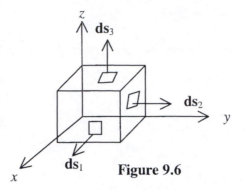

Figure 9.6

The surface integral of the vector **A** on the right hand side of preceding expression integrated over a surface is known as the *flux* of **A**. For example, if **A** represents the velocity **v** of a fluid, then the flux of **v** would mean the total amount of fluid passing the surface in unit time. Since divergence represents the spreading out of the vector, then, if the fluid is incompressible, the total amount of fluid produced or created by all sources is equal to the total amount of fluid passing through all the surfaces enclosing the volume. This is what is expressed by Gauss divergence theorem.

If for a vector $\displaystyle\int_{volume} (\nabla\cdot\mathbf{A})\,dV = 0$ (such as for the magnetic field **B**), then from the

divergence theorem, the flux $\displaystyle\oint_{surface} \mathbf{A}\cdot\mathbf{ds} = 0$. This physically means that as much flux

enters, the same amount of flux leaves across the surface bounding a given volume.

Example 9.3.2

Verify Gauss's divergence theorem for the vector $\mathbf{A} = c\mathbf{r}$ (where c is a constant and r is the position vector) and for a cube of side a centered at the origin.

First let us first find out $\displaystyle\int_{volume} (\nabla\cdot\mathbf{A})\,dV$.

$$\nabla\cdot(c\mathbf{r}) = \left(\frac{\partial}{\partial x}\mathbf{i} + \frac{\partial}{\partial y}\mathbf{j} + \frac{\partial}{\partial z}\mathbf{k}\right)\cdot c(x\mathbf{i} + y\mathbf{j} + z\mathbf{k})$$

$$= c\,(1 + 1 + 1) = 3c.$$

Therefore, $\displaystyle\int_{volume} (\nabla\cdot\mathbf{A})\,dV$

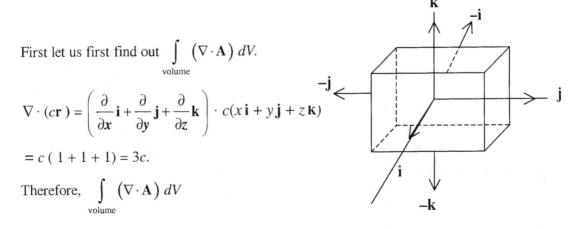

$$= \int_{volume} (3c)\, dV = 3c \int_{volume} dV = 3ca^3. \text{ (since the volume of the cube} = a^3)$$

Next determine $\oint_{surface} \mathbf{A} \cdot \mathbf{ds}$.

On each face of the cube, the direction of $\mathbf{ds}$ is $\pm\mathbf{i}$, $\pm\mathbf{j}$, or $\pm\mathbf{k}$, as shown.

Thus, the value of the integral for the face at $x = a/2$ is

$$\int_{surface} \mathbf{A} \cdot \mathbf{ds} = \int_{surface} c(x\,\mathbf{i} + y\,\mathbf{j} + z\,\mathbf{k}) \cdot \mathbf{i}\, ds.$$

Substitute $x = a/2$ and note that $\mathbf{i} \cdot \mathbf{j} = 0 = \mathbf{i} \cdot \mathbf{k}$. $\quad = \int_{surface} c(a/2)\, (\mathbf{i} \cdot \mathbf{i})\, ds$

Area of the surface $= a^2$ and $\mathbf{i} \cdot \mathbf{i} = 1$. $\qquad = c\left(\dfrac{a}{2}\right) a^2 = \dfrac{1}{2} ca^3.$

Similarly, the value of the integral for the face at $x = -a/2$ is

$$\int_{surface} \mathbf{A} \cdot \mathbf{ds} = \int_{surface} c(x\,\mathbf{i} + y\,\mathbf{j} + z\,\mathbf{k}) \cdot (-\mathbf{i})\, ds = \int_{surface} c(-a/2)\, (-\mathbf{i} \cdot \mathbf{i})\, ds = c\left(\dfrac{a}{2}\right) a^2 = \dfrac{1}{2} ca^3.$$

The result of the integral $\int_{surface} \mathbf{A} \cdot \mathbf{ds}$ is the same for all six surfaces.

Thus, for all the surfaces, $\oint_{surface} \mathbf{A} \cdot \mathbf{ds} = 6 \cdot \dfrac{1}{2} ca^3 = 3ca^3$. This verifies Gauss's

divergence theorem.

*Stokes's Theorem

Stokes's theorem states that the surface integral of the curl of a vector over a surface is equal to the line integral of the vector across the perimeter of the surface. That is,

$$\int_{surface} (\nabla \times \mathbf{A}) \cdot \mathbf{ds} = \oint_{loop} \mathbf{A} \cdot \mathbf{dr}.$$

The curl of a vector $\mathbf{A}$ measures the "curl" or the "twist" of the vector $\mathbf{A}$ at each point. Then the integral on the left-hand side of the preceding expression represents the total amount of curl or twist over a whole surface. However, we can determine the twisting motion also by moving around the edge and determining how much the flow is following the line forming the boundary of the surface. This is what is expressed in the right hand side of the preceding expression. This is represented pictorially below:

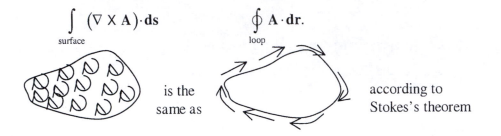

$$\int_{\text{surface}} (\nabla \times \mathbf{A}) \cdot \mathbf{ds} \qquad\qquad \oint_{\text{loop}} \mathbf{A} \cdot \mathbf{dr}.$$

is the same as   according to Stokes's theorem

Note that Stokes's theorem is analogous to the Gauss divergence theorem in a smaller number of dimensions since it involves a relation between a surface integral and a line integral.

Example 9.3.3

Verify Stokes's theorem for the vector $\mathbf{A} = (zx + 2y^2)\,\mathbf{j} + yz\,\mathbf{k}$ and for the square surface of length a as shown.

First, find the surface integral $\displaystyle\int_{\text{surface}} (\nabla \times \mathbf{A}) \cdot \mathbf{ds}.$

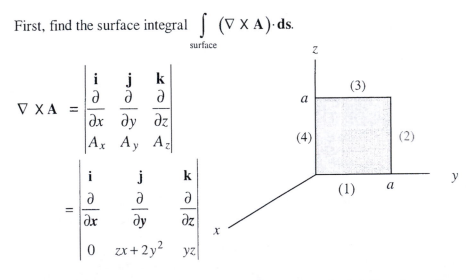

$$\nabla \times \mathbf{A} = \begin{vmatrix} \mathbf{i} & \mathbf{j} & \mathbf{k} \\ \dfrac{\partial}{\partial x} & \dfrac{\partial}{\partial y} & \dfrac{\partial}{\partial z} \\ A_x & A_y & A_z \end{vmatrix}$$

$$= \begin{vmatrix} \mathbf{i} & \mathbf{j} & \mathbf{k} \\ \dfrac{\partial}{\partial x} & \dfrac{\partial}{\partial y} & \dfrac{\partial}{\partial z} \\ 0 & zx+2y^2 & yz \end{vmatrix}$$

$= (z - x)\,\mathbf{i} + 0\,\mathbf{j} + z\,\mathbf{k} = (z - x)\,\mathbf{i} + z\,\mathbf{k}.$ $\mathbf{ds} = dy\,dz\,\mathbf{i}.$ For the surface $x = 0.$

Therefore, $\displaystyle\int_{\text{surface}} (\nabla \times \mathbf{A}) \cdot \mathbf{ds} = \int_0^a\int_0^a [z\mathbf{i} + z\mathbf{k}] \cdot dy\,dz\,\mathbf{i} = \int_0^a\int_0^a z\,dy\,dz = \int_0^a z\,dz\int_0^a dy = \frac{1}{2}a^3.$

Next find out the line integral $\oint\limits_{loop} \mathbf{A}\cdot\mathbf{dr}$. The integral needs to be determined by breaking it into four parts from (1) to (4). Remember, $x = 0$ for the surface.

1. $z = 0$. y changes from 0 to a, $\int\limits_{line} \mathbf{A}\cdot\mathbf{dr} = \int\limits_0^a \left(2y^2\mathbf{j} + yz\mathbf{k}\right)\cdot dy\,\mathbf{j} = \int\limits_0^a 2y^2\,dy = \dfrac{2}{3}a^3$.

2. $y = a$, z changes from 0 to a, $\int\limits_{line} \mathbf{A}\cdot\mathbf{dr} = \int\limits_0^a \left(2y^2\mathbf{j} + yz\mathbf{k}\right)\cdot dz\,\mathbf{k} = a\int\limits_0^a z\,dz = \dfrac{1}{2}a^3$.

3. $z = a$, y changes from a to 0, $\int\limits_{line} \mathbf{A}\cdot\mathbf{dr} = \int\limits_a^0 \left(2y^2\mathbf{j} + yz\mathbf{k}\right)\cdot dy\,\mathbf{j} = \int\limits_a^0 2y^2\,dy = -\dfrac{2}{3}a^3$.

4. $y = 0$, z changes from a to 0, $\int\limits_{line} \mathbf{A}\cdot\mathbf{dr} = \int\limits_a^0 \left(2y^2\mathbf{j} + yz\mathbf{k}\right) dz\,\mathbf{k} = \int\limits_a^0 yz\,dz = 0$

Thus, $\oint\limits_{loop} \mathbf{A}\cdot\mathbf{dr} = \dfrac{2}{3}a^3 + \dfrac{1}{2}a^3 - \dfrac{2}{3}a^3 + 0 = \dfrac{1}{2}a^3$.

Hence, $\int\limits_{surface} \left(\nabla\times\mathbf{A}\right)\cdot\mathbf{ds} = \oint\limits_{loop} \mathbf{A}\cdot\mathbf{dr}$, as expected.

Exercises

15. Verify Gauss's divergence theorem for the vector field $\mathbf{A} = y^2\,\mathbf{i} + xy\,\mathbf{j} + yz\,\mathbf{k}$ and a cube of side a as shown.

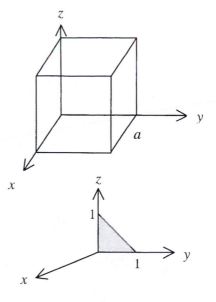

16. Verify Stokes's theorem for the vector function $\mathbf{A} = xy\,\mathbf{i} + yz\,\mathbf{j} + zx\,\mathbf{k}$ for the shaded triangular area.

17. Show using a diagram that the line integral about any surface is equal to the sum of all the line integrals about the parts into which the surface is divided.

Answers to Exercises in Chapter 9

1. (a) $\mathbf{i} - 2\mathbf{j} + 4\mathbf{k}$

    (b) $y\mathbf{i} + z\mathbf{j} + x\mathbf{k}$

    (c) $2\mathbf{r}$

4. (a) 0

    (b) 0

5. (a) $x + y + z$

    (b) x

7. (a) $-(y\mathbf{i} + z\mathbf{j} + x\mathbf{k})$

    (b) $2x\mathbf{i} + z\mathbf{j} + 2z\mathbf{k}$

11. $-\dfrac{1}{3}$

12. $0.02 \ \text{T·m}^2$

Chapter 10
How to Be Successful in a Physics Laboratory: Data Analysis, Curve Fitting, Probability, and Statistics

Statistical measurements are basically counts of the frequency of occurrence for different events in an experiment. Statistical analysis involves organizing the count data in the form of tables, graphing the data, and using the results to predict the future results. Data analysis is an important activity in statistics and in a physics laboratory. This chapter begins with some tips for success in a physics laboratory.

Experimental Error and Error Analysis

10.1   Introduction and Tips for Success in a Physics Laboratory

A physics laboratory is your chance to get hands-on experience and visually observe how physical principles are applied to real-world situations. The following tips should help you to enjoy your laboratory activities and improve your appreciation of laboratory physics:

- Be prepared before you come to the laboratory.

- Collaborate with other students in the group as well as other students in the lab.

- Think critically when you make any predictions or observations.

- Follow carefully the operational instructions and computer setup, if any.

- Take careful measurements.

- Analyze the data for the results. The analysis may involve making graphs or histograms or fitting curves (explained later in the chapter).

- Examine the results to draw a conclusion.

- Find experimental errors; that is, find the accuracy and the precision (explained later in the chapter) of your result wherever possible.

10.2   Experimental Data

In a physics laboratory, you will set up experiments, take measurements, collect data, and analyze data to get results and conclusions. Experimental data are usually of two types:

- Analytical

- Statistical

When you measure a physical quantity as a function of one or more other known physical quantities, the collected data set is *analytical*. For example, if you record the position of a freely falling object as a function of time in an experiment, the collected data set is analytical. In an analytical data set, you usually use the data to discover any relationship between physical quantities (such as how the position of a freely falling object depends on time).

But, when you record the frequency of occurrence of different events in an experiment, the collected data set is *statistical*. For example, suppose in an experiment you want to determine the energy distribution of the particles recorded in the experiment. In the experiment, you set up a small energy window and record how many particles have energy within this window. Then you vary the window several times and, each time, determine the number again, thus generating the counts for the number of particles for different energy intervals. In this example, the collected data set is statistical.

10.3   Experimental Errors

Laboratory measurements of physical quantities always involve some experimental errors, regardless of your efforts to do experiments as carefully as possible. In other words, any real measurement will have some *error* or *uncertainty*. These errors could be any combination of the following:

- Human error

- Instrumental error

- Computational error

Human errors arise from personal bias or from carelessness in reading an instrument, in recording observations, or in mathematical calculations. *Instrumental error* can occur because of the bad calibration of an instrument or its limitations to record or measure accurately. *Computational errors* arise because no calculator or computer can calculate with infinite perfection and because of error propagation during different mathematical operations.

Experimental errors could be

- Systematic

- Random

Systematic errors are associated with particular instruments or techniques, such as an improperly calibrated instrument, incorrect use of the measuring device, or any bias on the part of the observer. One important source of systematic error is the *parallax*, which causes an apparent change in position due to the change in position of the eye—that is, looking at the measuring device from a wrong angle. For example, the length of the rod in Figure 10.1 may appear to be 8 or 10 units, respectively, if viewed above or below a line of sight perpendicular to the scale, although 9 units is the actual or true length. Systematic errors move the result in a single direction and cannot be minimized by repeating the measurements.

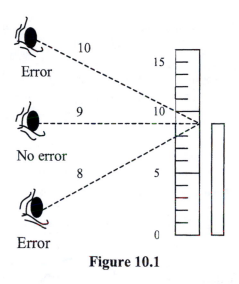

Figure 10.1

As another example of systematic error, think of an experiment that records the position of a freely falling object as a function of time and determines its velocity as a function of time and acceleration due to gravity *g* from the analysis of data. To determine *g*, it will be assumed that the velocity of the object increases linearly with time. The assumption holds for an ideal situation where friction and air drag are absent. In reality, the velocity would be less (although it could be negligibly small) because of these opposite effects. We can minimize systematic errors in an experiment by performing the experiment in a better environment (such as in a vacuum).

Random error results from unknown and unpredictable variations in experimental situations, such as temperature fluctuation of the environment. For random errors, the more measurements you make, the more errors will cancel each other. Thus, you can minimize random errors by repeating measurements many times and then taking the average of the measurements.

Since error cannot be totally avoided in a physics laboratory regardless of how good you are as an experimentalist, the analysis and reporting of errors is very important in a physics experiment.

10.4 Accuracy and Precision of Measurements: Mean or Average, Average Deviation, Variance, and Standard Deviation

You should understand the meaning of the terms *precision* and *accuracy* of a result that are commonly associated with the experimental measurements.

> **Note:** It is often said that *a measuring instrument that is high in precision is also high in accuracy.* This is not necessarily true! An instrument can provide precise results that could be way off from the actual result. For example, if you use an improperly calibrated ruler to measure a length carefully several times, your results can be precise but not accurate.

Precision means the agreement among repeated measurements, that is, how close the individual results are. It is a measure of the magnitude of uncertainty of the result, without any reference to what the actual result is. For example, say two independent experiments of a time measurement provide the following results: 1.2 ± 0.1 seconds and 1.3 ± 0.2 seconds, respectively. The spread in the first measurement is between 1.1 and 1.3 seconds, whereas the spread in the second measurement is 1.1 and 1.5 seconds. Since the spread is less in the first measurement, the first measurement is more precise than the second measurement.

The *accuracy* of a measurement tells us how close the result is to the true or the accepted value. It is a measure of correctness of a result. The accepted value of a physical quantity can be found in the *CRC Physics Handbooks* (CRC Press, New York) or in physics textbooks. For example, suppose two independent measurements for the determination of π provide the results 3.141 and 3.147, respectively. Since the true value of π up to four significant figures is 3.142, the first result is more accurate since it is closer to the true value.

To clarify further the basic difference between precision and accuracy, look at Figures 10.2a–d for an analogy, where the true value is the bull's eye. Figure 10.2a has poor accuracy and poor precision. Figure 10.2b has poor accuracy, but good precision. Figure 10.2c has good accuracy, but poor precision. Figure 10.2d has good accuracy and good precision.

Figure 10.2a–d Accuracy and Precision.

The *accuracy* of a measurement can be expressed by the *percentage error*:

$$\text{Percentage error} = \frac{|\text{Accepted Value - Measured Value}|}{\text{Accepted Value}} \times 100\% .$$

The *precision* of a measurement is expressed by calculating the *average deviation* or the *standard deviation* of the results.

The *average deviation* or *mean deviation* is the mean of absolute deviations of the individual results from the average value of the measurements. Thus, if $x_1, x_2, \ldots, x_N$ are N individual measurements and $\bar{x}$ is their average value, then

$$\bar{x} = \frac{x_1 + x_2 + \ldots + x_N}{N} .$$

Absolute deviation of $x_1 = |x_1 - \bar{x}|$,

Absolute deviation of $x_2 = |x_2 - \bar{x}|$, and so on.

Thus, the average deviation (a.d.) is given by

$$\text{a.d.} = \frac{|x_1 - \bar{x}| + |x_2 - \bar{x}| + \ldots + |x_N - \bar{x}|}{N}$$

$$= \frac{1}{N} \sum_{i=1}^{N} |x_i - \bar{x}| .$$

As the number of measurements is increased, the average of the set of measurements becomes more and more accurate in proportion to the square root of the number of measurements, N. This is reflected in the average deviation of the mean (A.D.) given by

$$\text{A.D} = \frac{\text{a.d.}}{\sqrt{N}} .$$

The A.D. represents the deviation of the arithmetical mean from the actual value and is known as the *probable error*. According to the theory of probability, A.D. represents a 50% chance that the actual value of the quantity would lie between the A.D. of the mean.

The average of the square of the deviations of a set of measurements is called the *variance*, σ^2; that is,

$$\sigma^2 = \frac{\left(x_1 - \bar{x}\right)^2 + \left(x_2 - \bar{x}\right)^2 + \dots + \left(x_N - \bar{x}\right)^2}{N}$$

$$= \frac{1}{N} \sum_{i=1}^{N} \left(x_i - \bar{x}\right)^2.$$

The *standard deviation* σ is defined as the square root of the variance, that is, the square root of the squares of the individual deviations. Thus,

$$\sigma = \sqrt{\frac{1}{N} \sum_{i=1}^{N} \left(x_i - \bar{x}\right)^2}.$$

From the standpoint of the statistical theory, the standard deviation is the most important measure of dispersion, that is, the scatter of experimental points. When the number of measurements is small, it can be shown statistically that a better estimate of standard deviation is given by

$$\sigma = \sqrt{\frac{1}{(N-1)} \sum_{i=1}^{N} \left(x_i - \bar{x}\right)^2}.$$

The standard deviation is also called the *root mean square deviation* (or rms deviation), because you take the square root of the average (mean) of the squares of the deviations.

The *standard deviation of the mean* (SDOM) is given by

$$\text{SDOM} = \frac{\sigma}{\sqrt{N}}.$$

It is useful to know that when a large number of measurements of a quantity are made, the results form a *normal distribution* or *Gaussian distribution*, $p(x)$, if the variations of the individual results occur due to random effects. A bell-shaped curve, symmetric about the average value of the measurements, represents this distribution, which is shown in Figure 10.3. In this case, it can be shown statistically that there is a 68% chance (that is, 68% probability, in the language of statistics) that a measurement will fall within one standard deviation of the average value, $\overline{x} - \sigma < x < \overline{x} + \sigma$. The chance, or probability, that a measurement will fall within the two standard deviations of the average, $\overline{x} - 2\sigma < x < \overline{x} + 2\sigma$, can be shown to be 95%. See Appendix K for more details about Gaussian distribution.

Gaussian distribution

Figure 10.3

Example 10.4.1

In a series of experimental measurement for the value of the wavelength of a helium-neon laser light, the following results (in nanometers, 1 nanometer = 10^{-9} m) are obtained: 636.1, 641.2, 631.3, 645.1, 628.4, 624.8, 628.4, 627.8, 625.7, and 644.1.

Find the average value, $\overline{x}$, of the wavelength.

Find the percentage error of the measurement, where the accepted value of the wavelength is 632.8 nanometers. Find the average deviation (a.d.) of the results. Find the variance and standard deviation σ of the results. Also, find the A.D. and SDOM.

In this case, the average, $\overline{x} = \dfrac{x_1 + x_2 + \ldots + x_N}{N}$.

The number of measurements $N = 10$. Therefore,

$$\overline{x} = \frac{636.1 + 641.2 + 631.3 + 645.1 + 628.4 + 624.8 + 628.4 + 627.8 + 625.7 + 644.1}{10} \text{ nm}$$

= 633.3 nm.

$$\text{Percentage error} = \frac{|\text{Accepted Value - Measured Value}|}{\text{Accepted Value}} \times 100\%$$

$$= \frac{|632.8 - 633.3|}{632.8} \times 100\% = 0.08\%$$

$$\text{Average deviation (a.d.)} = \frac{|x_1 - \bar{x}| + |x_2 - \bar{x}| + ... + |x_N - \bar{x}|}{N}$$

$$=$$

$$\frac{1}{10}[|(636.1 - 633.3)| + |(641.2 - 633.3)| + |(631.3 - 633.3)| + |(645.1 - 633.3)| +$$

$$|(628.4 - 633.3)| + |(624.8 - 633.3)| + |(628.4 - 633.3)| + |(627.8 - 633.3)| +$$

$$|(625.7 - 633.3)| + |(644.1 - 633.3)|]$$

$$= \frac{1}{10}[|2.8| + |7.9| + |-2.0| + |11.8| + |-4.9| + |-8.5| + |-4.9| + |-5.5| + |-7.6| + |10.8|]$$

$$= \frac{1}{10}[2.8 + 7.9 + 2.0 + 11.8 + 4.9 + 8.5 + 4.9 + 5.5 + 7.6. + 10.8]$$

= 6.7 nm

Thus, the wavelength = (633.3 ± 6.7) nm. That is, the actual value of the wavelength has a high probability of lying between 626.6 nm (633.3 − 6.7 nm) and 640.0 nm (633.3 + 6.7 nm). Statistically, it can be shown that when the number of measurements is large, 57.5% of the individual results would lie within this interval (a.d.).

$$\text{Variance, } \sigma^2 = \frac{(x_1 - \bar{x})^2 + (x_2 - \bar{x})^2 + ... + (x_N - \bar{x})^2}{N}$$

$$=$$

$$\frac{1}{10}\left[\begin{array}{l} (2.8)^2 + (7.9)^2 + (-2.0)^2 + (11.8)^2 + (-4.9)^2 + (-8.5)^2 + (-4.9)^2 + (-5.5)^2 + \\ (-7.6)^2 + (10.8)^2 \end{array} \right]$$

$= 53.8 \text{ nm}^2$

Standard deviation $\sigma = \sqrt{53.8} \text{ nm} = 7.3 \text{ nm}$

This means that there is a 68% chance that an individual measurement of the wavelength would lie within (633.3 ± 7.3) nm.

$$\text{A.D.} = \frac{\text{a.d.}}{\sqrt{N}} = \frac{6.7 \text{ nm}}{\sqrt{10}} = 2.1 \text{ nm}$$

This means that the chance that the true value of the wavelength lies within the interval (633.3 ± 2.1) nm is 50%.

$$\text{SDOM} = \frac{\sigma}{\sqrt{N}} = \frac{7.3 \text{ nm}}{\sqrt{10}} = 2.3 \text{ nm}$$

Exercise

1. In separate experimental measurements by ten different students in a class for the value of the acceleration due to gravity (g) at the same laboratory, the following results (in meters per second squared or m/s^2) are found: 9.82, 9.79, 9.68, 9.83, 9.81, 9.78, 9.72, 9.77, 9.84, and 9.86. Find the average value of g for the class. Find the percentage error of the average value if the accepted value of g at that place is 9.81 m/s^2. Find the average deviation (a.d.) and standard deviation σ of the result. Also, find the A.D. and SDOM.

10.5 Propagation of Error

Errors become larger, or, in other words, errors propagate when you combine different measurements and their errors. Therefore, you should pay attention when combining mathematically different measurements and their errors.

Let us call δx_1, δx_2 the errors (which may be the standard deviations of each measurement) for the measurements of the two variables, x_1, x_2, respectively. In other words, δx_1, δx_2 are the *uncertainties* of measurement x_1, x_2, respectively. Table 10.1 summarizes the results for the total error or total uncertainty in different mathematical operations of x_1, x_2. Use the appropriate formula from the table when you mathematically combine two different results with their individual errors. Note that the fractional error in x_1 is $\delta x_1/x_1$, and the percentage error in x_1 is $(\delta x_1/x_1)$ x 100%, and so on.

Table 10.1 Expressions for Total Uncertainty in Mathematical Operations

Mathematical Operation	Results with Total Uncertainty
Addition	$(x_1 \pm \delta x_1) + (x_2 \pm \delta x_2) = (x_1 + x_2) \pm \sqrt{\delta x_1^2 + \delta x_2^2}$
Subtraction	$(x_1 \pm \delta x_1) - (x_2 \pm \delta x_2) = (x_1 - x_2) \pm \sqrt{\delta x_1^2 + \delta x_2^2}$
Multiplication	$(x_1 \pm \delta x_1) \cdot (x_2 \pm \delta x_2) \approx (x_1 x_2) \pm$ $(x_1 x_2)\sqrt{\left(\dfrac{\delta x_1}{x_1}\right)^2 + \left(\dfrac{\delta x_2}{x_2}\right)^2}$
Division	$\dfrac{(x_1 \pm \delta x_1)}{(x_2 \pm \delta x_2)} \approx \dfrac{x_1}{x_2} \pm \dfrac{x_1}{x_2}\sqrt{\left(\dfrac{\delta x_1}{x_1}\right)^2 + \left(\dfrac{\delta x_2}{x_2}\right)^2}$
Multiplication by a constant	$A(x + \delta x) = Ax \pm A\delta x$
Raised to a power	$(x \pm \delta x)^r = x^r \pm rx^{r-1}\delta x$ (r any real number, not just integer)

Example 10.5.1

The lengths of two rods were measured and found to be 60.2 ± 0.1 cm and 64.1 ± 0.2 cm, respectively.

(a) When the rods are joined, find the length of the combined rod.
(b) A bug is found to crawl along the first rod starting from its one end with a speed of 2.1 ± 0.3 cm/s. How long does it take to reach the other end of the rod?
(c) Find the area of a circle with the first rod as its radius.
(d) Find the area of the rectangle whose two sides have the same lengths as the rods.

(a) To determine the length of the combined rod with uncertainty, use the formula

$$(x_1 \pm \delta x_1) + (x_2 \pm \delta x_2) = (x_1 + x_2) \pm \sqrt{\delta x_1^2 + \delta x_2^2}.$$

Therefore, the length of the combined rod $= (60.2 + 64.1) \pm \sqrt{(0.1)^2 + (0.2)^2}$

Simplify the result. $= (124.3 \pm 0.2)$ cm.

(b) To determine the time, we divide the length x by the speed v.

Therefore,   time $= \dfrac{(x \pm \delta x)}{(v \pm \delta v)} = \dfrac{x}{v} \pm \dfrac{x}{v} \sqrt{\left(\dfrac{\delta x}{x}\right)^2 + \left(\dfrac{\delta v}{v}\right)^2}.$

Substitute the values. $= \dfrac{60.2}{2.1} \pm \dfrac{60.2}{2.1} \sqrt{\left(\dfrac{0.1}{60.2}\right)^2 + \left(\dfrac{0.3}{2.1}\right)^2}$

Simplify. $= (28.67 \pm 4.1)$ seconds

Round off the result. $= (29 \pm 4)$ seconds

(c) The area of the circle $A = \pi r^2 = \pi (60.2)^2 = 11400$ cm^2.

The uncertainty is given by $\dfrac{\delta A}{A} = 2\dfrac{\delta r}{r}$ [since, if $y = ax^n$, then $\dfrac{\delta y}{y} = n\dfrac{\delta x}{x}$].

$$\delta A = 2\dfrac{\delta r \cdot A}{r} = 2\dfrac{0.1 \times 11379.49}{60.2} = 37.8 \text{ cm}^2$$

Thus, area $= (11379.49 \pm 37.8)$ cm^2

Round off the result. $= (11380 \pm 40)$ cm^2

(d) The area of the rectangle $=$

$$(x_1 \pm \delta x_1) \cdot (x_2 \pm \delta x_2) = (x_1 x_2) \pm (x_1 x_2)\sqrt{\left(\dfrac{\delta x_1}{x_1}\right)^2 + \left(\dfrac{\delta x_2}{x_2}\right)^2}.$$

Substitute the values. $= (60.2)(64.1) \pm (60.2)(64.1)\sqrt{\left(\dfrac{0.1}{60.2}\right)^2 + \left(\dfrac{0.2}{64.1}\right)^2}$

Simplify. $= (3858.82 \pm 13.64)$ cm^2

Round off the result. $= (3859 \pm 14)$ cm^2

Exercise

2. The lengths of two rods were measured and found to be 30.1 ± 0.3 cm and 10.2 ± 0.2 cm, respectively. (a) When the rods are joined, find the length of the combined rod; (b) find the area of a rectangle with first rod as the base and the second rod as the height of the triangle; (c) find the area of the circle with the first rod as its radius and (d) determine the ratio of the lengths of the two rods.

Plots and Curve Fitting

10.6 Analysis and Plot of Data

In an experiment in physics, measurements are made of the variable(s) of interest, and the raw result of an experiment is a set of measurements, called the *data set*. Usually in an experiment, there are two kinds of data. The quantity that is varied in the experiment is called the *independent variable*, and the quantity that is measured is called the *dependent variable*. The collected data sets are then analyzed correctly to get the final result. The analysis usually includes organizing the data set in the form of a table or spreadsheet, making a graph, fitting the data to an appropriate equation or function, and determining the final result and errors in the measurements.

A *table* is a listing of variables with each variable in separate columns. In most situations, the independent variable is arranged, in increasing order of its magnitudes, in the first column. When you make a table, it should have an appropriate title, and the column heads must have the names of the variables with their proper units.

Note: A common mistake made by students is to write only the names of the variables in the column heads of a table without their units. Make sure to enter the proper units in parentheses as shown here, where "s" stands for seconds and "m" for meters:

Time (s)	Position (m)
1	4.9
2	19.5, etc.

In a *Cartesian graph*, you will plot the variables, usually with the independent variable along the x-axis and the dependent variable along the y-axis. Make sure to give a *title* to the graph and *label the variables with their units*. The scale of the axes should be such that the graph fills the entire sheet of a paper or a reasonable portion of it, as appropriate. For preliminary inspection, a scattered graph is usually all right, but for the final graph, connect the points by a smooth line. If any experimental error for any data point needs to be indicated on the graph, indicate the errors using proportionately sized vertical lines drawn at the points. These proportionately sized vertical lines drawn at different points on a graph are known as the *error bars*. Figure 10.4a shows a scattered plot of y versus x with error bars. There may be situations where you need to extend the graph beyond the range of the collected data set; that is, you need to *extrapolate* the graph to find the result. It is common to extrapolate a graph so that it meets the y-axis to get the y intercept at $x = 0$, the value of which may be physically meaningful. Figure 10.4b shows that the straight line has been extrapolated (shown as a broken line) to meet the y-axis. Similarly, in some situations where data sets have missing data, you need to guess the trend; that is, you need to *interpolate* the graph for the result. Figure 10.4c shows that the data have been interpolated (broken line) to complete the line, indicating how y changes with x for the whole range of x.

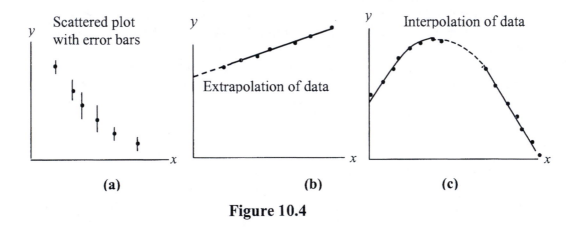

Figure 10.4

Note: There is always a risk involved in extrapolating and interpolating. For example, when you measure the resistance of a material and find that the resistance decreases linearly, if you extrapolate the line back to zero resistance, you may miss a critical temperature when the material becomes superconductor. Also, if you measure the sound intensity as a function of frequency, if you interpolate the data over a wide frequency gap, you may miss a resonant peak.

10.7   Semilog and Log-Log Graphs

A picture is worth a thousand words, and a well-constructed graph is worth at least five thousand, especially when it displays a large amount of information in a small space to make the patterns visible. Several coordinate systems and graph paper have been devised for graphical analysis and plotting. In introductory physics, you will use mostly *rectangular*, or *Cartesian*, graphs, which are described in Section 4.4. On some occasions, you will need to plot logarithms of data. In a rectangular Cartesian grid, the tick marks are equally spaced for both the *x*- and the *y*-axes. If the *grid is logarithmic*, the tick marks are not equally spaced. The spaces for numbers beginning with 1 and 2 are larger, compared with the numbers beginning with 8 or 9. There are two types of logarithmic graphs:

- Semilog graph

- Log-log graph

In semilog graphs, only the *y*-axis is logarithmic (i.e., is, separation between ticks is proportional to the logarithm of numbers), while for the *x*-axis the ticks are equally spaced, as with for a rectangular Cartesian graph. For log-log graphs both the *x*- and the *y*-axes are logarithmic. Semilog and log-log grids are shown in Figures 10.5a and b, respectively.

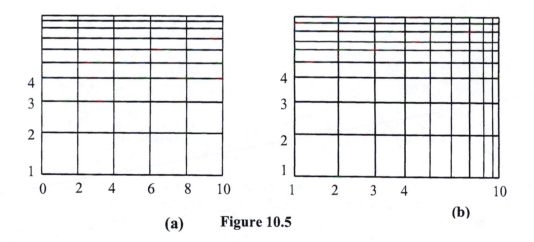

(a)   **Figure 10.5**   (b)

Semilog and log-log graphs have advantages over Cartesian graphs in some situations. For example, if the relationship between two physical quantities is $y = A\,e^{ax}$, a plot of *x* and *y* on a regular Cartesian graph gives an exponential curve, which is nonlinear. Taking the logarithm of the preceding equation, we get

$$\ln y = ax + \ln A.$$

Thus, a plot of $\ln y$ (or $\log y$, since $\ln y \approx 2.3 \log y$) versus x on a Cartesian graph, or, in other words, a plot of x and y on a semilog paper, gives a straight line with slope a and intercept $\ln A$. For this reason, when the relationship between two physical quantities is exponential, a semilog graph is often used.

A plot $y = 2\,e^{3x}$ is shown in the Cartesian graph in Figure 10.6. Figure 10.7 shows the Cartesian plot of $\ln y$ versus x for the equation $y = 2\,e^{3x}$. Note that the plot of $\ln y$ versus x is a straight line with intercept $\ln 4$ ($= 0.60$) and slope 3. (Verify this by considering any two points on the graph!)

$$y = 2\,e^{3x}$$

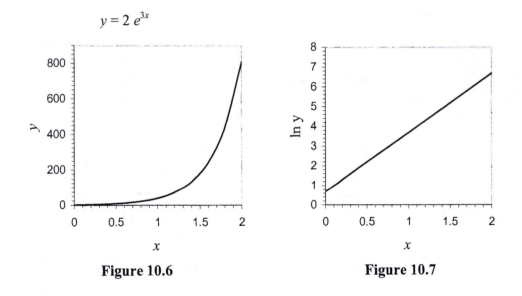

Figure 10.6   **Figure 10.7**

If the relationship between two physical quantities x and y is a power function, $y = ax^n$ (where n is an integer other than 0 and 1), the graph is clearly a nonlinear curve in the Cartesian plot. If we take the logarithm of the power equation, $y = ax^n$, we get

$$\ln y = \ln a + n \ln x.$$

Thus, a plot of $\log y$ versus $\log x$, or, in other words, a log-log plot of x and y will produce a straight line with slope n and y-intercept $\log a$. From the knowledge of the slope of the straight line, it is easy to determine the value of a and n. For this reason, a log-log graph is often used when the relationship is a power function.

A plot of $y = 4x^3$ is shown in the Cartesian graph in Figure 10.8. Figure 10.9 shows the corresponding log-log plot. Note that the plot is a straight line with intercept ln 4 and slope 3. (Verify this by considering any two points on the graph!)

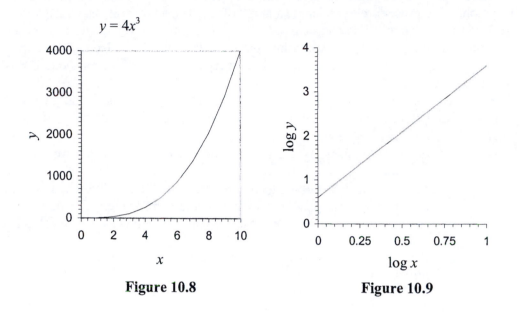

Figure 10.8   **Figure 10.9**

Exercise

3. Use the accompanying data to make a Cartesian plot and show that it is a nonlinear curve passing through the origin. Then, make a log-log plot, and show that it is a straight line. From the plots, discover that the relationship between the diameter D and volume V of a sphere is $V = 0.52\,D^3$. The data are as follows:

Diameter (m)	Volume (m³)
0.10	0.00052
0.15	0.00177
0.20	0.00419
0.25	0.00818

10.8 Curve Fitting

In many situations, it is necessary to examine the plot to determine the mathematical function that describes the relationship between the variables plotted on the graph. The procedure of matching a graph of experimental variables with a known mathematical function is known as *curve fitting*. In curve fitting, you determine the best function that describes the relationship between the variables. Note that due to experimental error it is likely that the function may not pass through all the points of the graph.

In most situations in introductory physics, the fitted function is a straight line, a polynomial function of the independent variable, an exponential function, or a power function:

- A *linear* relationship between two variables x and y can be expressed by the equation $y = mx + b$, where m is called the slope of the line, defined in Chapter 4, and b is the y-intercept of the line (the y value where $x = 0$). Thus, the relationship is known when m and b are known. The purpose of the curve fitting is to determine the constant m and b if the relationship is a straight line.

- A *polynomial* function relationship between x and y is expressed by an equation $y = a_0 + a_1x + a_2x^2 + a_3x^3 + \cdots$, where $a_0, a_1, a_2, a_3, ..$ are constants to be determined from the curve fitting.

- An *exponential* relationship is expressed by $y = ae^{bx}$, where a and b are constants, to be determined from the curve fitting. Note that if you take the logarithm of the equation, it gives $\ln y = bx + \ln a$, which is a straight line if $\ln y$ is plotted along the y-axis with x along the x-axis.

- For a *power function*, the relationship is $y = ax^n$, where n is a positive integer and a is a constant to be determined from the curve fitting. If you take the logarithm on both sides, $\ln y = n \ln x + \ln a$. Thus, a plot of $\ln y$ versus $\ln x$ is a straight line.

Note: If the relationship is exponential, make a semilog plot, and if the relationship is a power function, make a log-log plot to find the coefficients.

The theory of curve fitting is not simple. The kind of mathematics and the level of difficulty depend on the shape of the fitted curve. *Least squares fitting* fits data to a curve or a line so that the *square of the deviations of the data points from the fitted curve or line is the least*. If the data seem to fit to a straight line, least squares fitting, also called *linear regression*, is commonly used for the best-fitted line.

Least Square Fit to a Straight Line

Straight-line curve fitting is an important case and is used frequently since many sets of experimental data that can be fitted by a straight line or at least should theoretically be fitted by a straight line are collected in a physics laboratory.

Suppose we have N sets of data such as (x_1, y_1), (x_2, y_2), $\cdots$, (x_N, y_N) from N measurements. Least squares fit of the data to a straight line from the data is done by determining the constants m and b for the best-fitted line $y = mx + b$ by minimizing the expression for the sum of the square of the individual deviations, given by

$$\sum \left(y_i - \left(m x_i + b \right) \right)^2.$$

Let $D = \sum \left(y_i - \left(m x_i + b \right) \right)^2.$

For D to be minimum, we set $\dfrac{dD}{da} = 0$ and $\dfrac{dD}{dm} = 0.$

From $\dfrac{dD}{da} = 0$ we get $\displaystyle\sum_{i=1}^{N} y_i - \left(m \sum_{i=1}^{N} x_i + Nb \right) = 0$ or $\displaystyle\sum_{i=1}^{N} y_i = m \sum_{i=1}^{N} x_i + Nb.$

From $\dfrac{dD}{dm} = 0$ we get $\displaystyle\sum_{i=1}^{N} x_i y_i - \left(m \sum_{i=1}^{N} x_i^2 + b \sum_{i=1}^{N} x_i \right)$ or $\displaystyle\sum_{i=1}^{N} x_i y_i = m \sum_{i=1}^{N} x_i^2 + b \sum_{i=1}^{N} x_i.$

The preceding two equations at the right above contain two variables m and b and can be solved. It can be shown that the expressions of m and b are

$$m = \frac{N \sum x_i y_i - \sum x_i \sum y_i}{N \sum x_i^2 - \left(\sum_i x_i \right)^2}$$

and

$$b = \frac{\sum x_i^2 \sum y_i - \sum x_i \sum_i \left(x_i y_i \right)}{N \sum x_i^2 - \left(\sum_i x_i \right)^2}, \text{ where each summation is from } i = 1, 2,$$

$3, \cdots, N.$

You will generally be able to do the regression analyses of the experimental data using a standard calculator, computer spreadsheet, and other mathematics or physics software that you may use in your physics laboratory.

Example 10.8.1

A ball is dropped from a certain height. The accompanying data set represents the speed of the falling ball at different times:

Time (s)	Speed (m/s)
0.0	0.0
1.0	9.6
2.0	19.8
3.0	30.1
4.0	38.9
5.0	49.1
6.0	59.1
7.0	68.8

Make a plot of the best-fitted straight line from the data and determine the slope m and the y-intercept of the line.

The data are plotted to make a Cartesian graph of the speed of the object as a function of time. The best-fitted straight line that shows that the speed of the object increases linearly with time is shown in the following diagram:

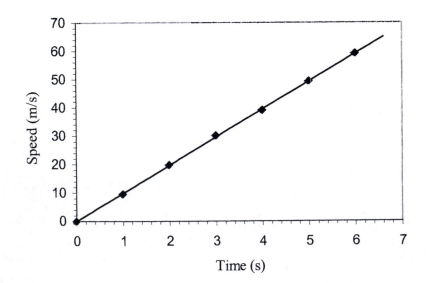

To determine the best-fitted line via linear regression, use the following expressions for the slope m and y intercept b:

$$m = \frac{N \sum x_i y_i - \sum x_i \sum y_i}{N \sum x_i^2 - \left(\sum_i x_i\right)^2}$$

$$b = \frac{\sum x_i^2 \sum y_i - \sum x_i \sum_i (x_i y_i)}{N \sum x_i^2 - \left(\sum_i x_i\right)^2}.$$

Here N (the number of measurements) = 8. Thus,

$\Sigma x_i = (0 + 1 + 2 + 3 + 4 + 5 + 6 + 7) \, s = 28 \, s$

$\Sigma x_i^2 = (0^2 + 1^2 + 2^2 + 3^2 + 4^2 + 5^2 + 6^2 + 7^2) \, s^2 = 140 \, s^2$

$\Sigma y_i = (0 + 9.6 + 19.8 + 30.1 + 38.9 + 49.1 + 59.1 + 68.8) \, m/s = 275.4 \, m/s$

$\Sigma x_i y_i = (0 \cdot 0 + 1 \cdot 9.6 + 2 \cdot 19.8 + 3 \cdot 30.1 + 4 \cdot 3.9 + 5 \cdot 49.1 + 6 \cdot 59.1 + 7 \cdot 68.8) \, m$

$\qquad = 1376.8 \, m$

$$m = \frac{N \sum x_i y_i - \sum x_i \sum y_i}{N \sum x_i^2 - \left(\sum_i x_i\right)^2} = \frac{8(1376.8 \, m) - (28 \, s)(275.4 \, m/s)}{8 \times 140 \, s^2 - (28 \, s)^2} = \frac{3303.2}{336} \, m/s^2$$

$= 9.8 \, m/s^2$

$$b = \frac{\sum x_i^2 \sum y_i - \sum x_i \sum_i (x_i y_i)}{N \sum x_i^2 - \left(\sum_i x_i\right)^2} = \frac{140 \, s^2 (275.4 \, m/s) - (28 \, s)(1376.8 \, m)}{8 \times 140 \, s^2 - (28 \, s)^2} = \frac{5.6}{336} \, m/s$$

$= 0.02 \, m/s = 0.0 \, m/s$ (after rounding off to tenth's place).

The relationship is $y = (9.8 \, m/s^2) \, x$.

Example 10.8.2

A mass attached to a light string suspended from a fixed support constitutes a simple pendulum. The time to complete one cycle of oscillation for the mass is called its period. In a physics laboratory, the period of a simple pendulum is recorded for different lengths of the pendulum. The recorded data are as follows:

Length (m)	Period (s)
0.125	0.74
0.225	0.98
0.325	1.14
0.425	1.36
0.525	1.44
0.625	1.60
0.725	1.75
0.825	1.82
0.925	1.95
1.025	2.06

Make a scatter plot of the length of the pendulum along the x-axis and its period along the y-axis, and show that the relationship between the length and period for a simple pendulum is nonlinear. Then plot the length with the square of its period and show that the relationship is linear. Do the regression analysis for the linear graph, and determine the slope and y-intercept of the line.

Following is the scatter plot of the data:

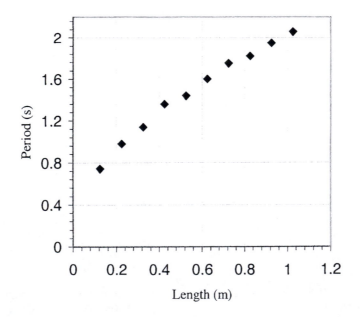

The relationship seems to be nonlinear.

From the raw data, square values of the periods are calculated, and the raw data, together with the square of the periods, are listed as follows:

Length (m)	Period (s)	Square of the Period (s^2)
0.125	0.74	0.55
0.225	0.98	0.96
0.325	1.14	1.30
0.425	1.36	1.85
0.525	1.44	2.07
0.625	1.60	2.56
0.725	1.75	3.06
0.825	1.82	3.31
0.925	1.95	3.80
1.025	2.06	4.24

Following is the best-fitted solid straight line for the plot of length versus square of the period:

The slope and *y*-intercept of the line are calculated from the regression analysis.

Here, N (the number of measurements) = 10.

Σx_i = 0.125 + 0.225 + 0.325 + 0.425 + 0.525 + 0.625 + 0.725 + 0.825 + 0.925 +

1.025 = 5.75 m

$\Sigma x_i^2 = (0.125)^2 + (0.225)^2 + (0.325)^2 + (0.425)^2 + (0.525)^2 + (0.625)^2 + (0.725)^2$
$+ (0.825)^2 + (0.925)^2 + (1.025)^2 = 4.13 \text{ m}^2$

$\Sigma y_i = 0.55 + 0.96 + 1.30 + 1.85 + 2.07 + 2.56 + 3.06 + 3.31 + 3.80 + 4.24 = 23.7 \text{ s}^2$

$\Sigma x_i y_i = 0.125 \cdot 0.55 + 0.225 \cdot 0.96 + 0.325 \cdot 1.30 + 0.425 \cdot 1.85 + 0.525 \cdot 2.07 +$
$0.625 \cdot 2.56 + 0.725 \cdot 3.06 + 0.825 \cdot 3.31 + 0.925 \cdot 3.80 + 1.025 \cdot 4.24$
$= 16.99 \text{ s}^2 \cdot \text{m}$

$$m = \frac{N \sum x_i y_i - \sum x_i \sum y_i}{N \sum x_i^2 - \left(\sum_i x_i \right)^2} = \frac{10\left(16.99 \text{s}^2 \text{ m} \right) - (5.75 \text{ m})\left(23.7 \text{s}^2 \right)}{10 \times 4.13 \text{ m}^2 - (5.75 \text{ m})^2} = \frac{33.62}{8.23} \text{ s}^2/\text{m}$$

$= 4.08 \text{ s}^2/\text{m}$

$$b = \frac{\sum x_i^2 \sum y_i - \sum_i x_i \sum (x_i y_i)}{N \sum x_i^2 - \left(\sum_i x_i \right)^2} = \frac{4.13 \text{ m}^2 \left(23.7 \text{s}^2 \right) - (5.75 \text{ m})\left(16.99 \text{ s}^2 \text{m} \right)}{10 \times 4.13 \text{ m}^2 - (5.75 \text{ m})^2} = \frac{0.19}{8.24} \text{ s}^2$$

$= 0.02 \text{ s}^2$

Therefore, the relationship is $T^2 = (4.08 \text{ s}^2/\text{m}) L + 0.02$, where L and T represent the length and period of the pendulum, respectively. Since 0.02 s^2 is negligible compared to the first term $(4.08 \text{ s}^2\text{m})L$ for the typical values of L (~1 m) , the equation reduces to

$T^2 = (4.08 \text{ s}^2/\text{m}) L$

or $T = \left(2.02 \text{ s}/\sqrt{\text{m}} \right)\sqrt{L}.$

Exercises

4. For the downhill motion of a glider on an inclined frictionless air track, the accompanying data were recorded for the position of the glider at different time intervals. The times were recorded by a PASCO tic-timer, which put tic marks on a tape moving with the glider at 0.1-second intervals.

Time (s)	Position (m)
0.0	0.000
0.1	0.021
0.2	0.045
0.3	0.075
0.4	0.106
0.5	0.137
0.6	0.170
0.7	0.207
0.8	0.245
0.9	0.286
1.0	0.330

Analyze the data to show that the relationship between the time t and the postion x of the glider is nonlinear and can be expressed by a quadratic equation $x = at + bt^2$. Determine the constant a and b.

5. In a physics laboratory a student shot a ball horizontally from a projectile launcher with a constant initial velocity. The student recorded the distance the ball travels in the x-direction, the distance the ball travels in the y-direction, and the total time t the ball in the air for four different heights of the projectile. The accompanying data set represent the data collected by the student:

Time, t (s)	Horizontal distance, x (m)	Vertical distance, y (m)
0.217	0.642	0.260
0.376	1.115	0.685
0.398	1.140	0.800
0.431	1.300	0.915
0.478	1.420	1.150
0.491	1.480	1.200

Analyze the data to determine the best-fitted curve for the motion of the projectile showing that the motion in the horizontal direction can be expressed by the equation $x = 3.00t$ while the motion in the y-direction can be best described by the equation $y = 4.95t^2$. Show also that the equation of motion of the projectile can be expressed by a parabola whose equation is $y = 0.55x^2$.

Probability and Statistics

10.9 Probability

The concept of probability is important in physics and mathematics, and especially in statistical physics. Introductory wave mechanics, which represents the behavior of subatomic particles such as electrons and protons, treats the wave equation as an expression for the mathematical probability.

The probability of an occurrence tells us the chance or likelihood of the occurrence. For example, when we toss a coin, there are two possible outcomes: Heads (H) and Tails (T). If we toss the coin a large number of times, it is found that the chance, or the probability, of obtaining H or T is ½.

If an event occurs in a ways and fails to occur in b ways, and each of these events is equally likely, then the probability of occurring the event is $a/(a + b)$.

Thus, the mathematical probability is defined as the *ratio of a certain number of events to the total number of events*.

The mathematical probability is a fraction that is always less than or equal to unity. When the probability attains its maximum value of unity, it becomes a certainty. By definition, the total probability for obtaining all states is unity. Probability is expressed as a fraction, decimal, or percentage. A probability of ½ can be written as ½, 0.5, or 50%.

Example 10.9.1

Two coins are simultaneously tossed on a table. Make a list of the different possible outcomes and the probability of each outcome.

The possible outcomes and their numbers are listed in the following table:

Number of Heads	Possible Outcomes	Number of Outcomes
2	HH	1
1	HT, TH	2
0	TT	1

Total number of possible outcomes = 1 + 2 + 1 = 4.

Thus,

the probability of 2 heads $= \dfrac{1}{4}$;

the probability of 1 head $= \dfrac{2}{4} = \dfrac{1}{2}$; and

the probability of 0 head (2 tails) $= \dfrac{1}{4}$.

Note that total probability for all outcomes $= \dfrac{1}{4} + \dfrac{1}{2} + \dfrac{1}{4} = 1$, as expected.

Example 10.9.2

Four coins are simultaneously tossed on a table. Make a list of the different possible outcomes and the probability of each outcome.

The possible outcomes and their numbers are listed in the following table:

Number of Heads	Possible Outcomes	Number of Outcomes
4	HHHH	1
3	HHHT, HHTH, HTHH, THHH	4
2	HHTT, HTHT, HTTH, THTH, TTHH, THHT	6
1	HTTT, THTT, TTHT, TTTH	4
0	TTTT	1

Total number of possible outcomes $= 1 + 4 + 6 + 4 + 1 = 16$.

Thus,

the probability of 4 heads (that is, 0 tail) $= \dfrac{1}{16}$;

the probability of 3 heads (that is, 1 tail) $= \dfrac{4}{16} = \dfrac{1}{4}$;

the probability of 2 heads (that is, 2 tails) $= \dfrac{6}{16} = \dfrac{3}{8}$;

the probability of 1 head (3 tails) $= \dfrac{4}{16} = \dfrac{1}{4}$; and the probability of 0 head (that is, 4 tails) $= \dfrac{1}{16}$.

The total probability $= \dfrac{1}{16} + \dfrac{1}{4} + \dfrac{3}{8} + \dfrac{1}{4} + \dfrac{1}{16} = 1$, as expected.

Probability of a Composite Event

Multiply individual probabilities to obtain sequential probabilities. Note that when only one coin is tossed, the probability of obtaining a head is ½. As found in Example 6.8.1, when two coins are simultaneously tossed, the probability of obtaining two heads is ¼, which is ½ × ½, the product of individual probabilities of obtaining each head separately. Similarly, when four coins are simultaneously tossed, the probability of obtaining four heads is 1/16, which is ½ × ½ × ½ × ½.

Thus, if two events are mutually independent and P_1 is the probability of event 1 and P_2 is the probability of event 2, then the probability of both events occurring simultaneously is given by
$$P_{12} = P_1 \times P_2.$$

Exercise

6. Calculate the probabilities, when you simultaneously throw two dice, of obtaining (a) a 5 and (b) a 9. (*Hint*: There are 36 possible outcomes.)

7. Suppose you shake six coins in your hand and drop them on a table.
(a) What is the total number of possible outcomes?
(b) How many different ways can you have 5 heads and 1 tail?
(c) How many different ways can you have 4 heads and 2 tails?
(d) What is the probability for obtaining 6 heads?
(e) What is the probability for obtaining 5 heads and 1 tail?
(f) What is the probability for obtaining 4 heads and two tails?

10.10 Frequency Distribution, Histogram, and Probability Distribution

In a statistical measurement, the counts of certain events are recorded. Some events occur more frequently than others. In the language of statistics, this means that some events are more probable than others. You get a *frequency distribution* of different counts by plotting the number of counts for an event along the *y*-axis and the value of the events along the *x*-axis. When the frequency distribution is plotted in the form of a bar graph, it is called a *histogram*. In a histogram, the height of each bar in the graph is proportional to the frequency of events. The width of the bar is called the *bin*. Each bin in a histogram is the same so that we can compare the bars with each other.

It is common to plot a statistical data set in the form of a *probability distribution* instead of in the form of a frequency distribution or histogram. To get a probability distribution, the frequency of occurrence of an event or a count is first divided by the total number of events. If f is the frequency of a measurement and N is the total number of measurements, then the probability of occurrence of a measurement is $p = f/N$. In the graph, p is then plotted along the *y*-axis, the event along the *x*-axis. The resultant graph is usually shown as a histogram when the number of bins is not large. When the number of bins is large, usually the points are joined by a smooth line. This line represents the *limiting probability distribution* of the measurements of the events. The common probability distributions are given in Appendix K.

Example 10.10.1

In a get-together of boys and girls, the ages of 800 boys and girls were recorded in years and months. Make a histogram of the data.

The data consist of 800 different records. The *bin* size is chosen to be 1 year for this histogram; that is, the data were sorted so that ages between 10.5 years 11.4 years will fall into one category of 11 years, and so on. The frequency distribution and the corresponding histogram are as follows:

Age (years)	Frequency, f	Age (years)	Frequency, f
1	0	11	155
2	0	12	140
3	1	13	94
4	1	14	48
5	2	15	22
6	6	16	8
7	20	17	2
8	42	18	1
9	100	19	0
10	150	20	0

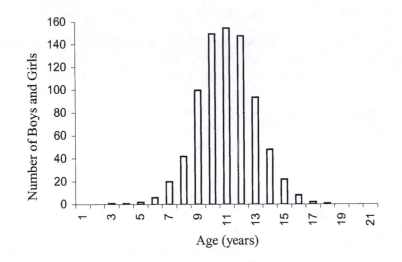

Example 10.10.2

For the previous example of a get-together of boys and girls, where the ages of 800 boys and girls were recorded in years and months, make a probability distribution of the data in form of a bar graph and a continuous line graph.

In general, to determine a probability distribution, you need to divide the frequency distribution by the total number of measurements or records. In this example, the total number N of boys and girls = 800. For example, the probability that a boy or girl will be age 10 is $p = f/N = 150/800 = 0.1875$. In a similar manner, determine the probability for each age. The resulting probability distribution data are as follows:

Age (years)	Probability, p	Age (years)	Probability, p
1	0	11	0.19375
2	0	12	0.185
3	0.00125	13	0.1175
4	0.00125	14	0.06
5	0.0025	15	0.0275
6	0.0075	16	0.01
7	0.025	17	0.0025
8	0.0525	18	0.00125
9	0.125	19	0
10	0.1875	20	0

The data are plotted in the form of a bar graph with the x-axis as age in years and the y-axis as probability:

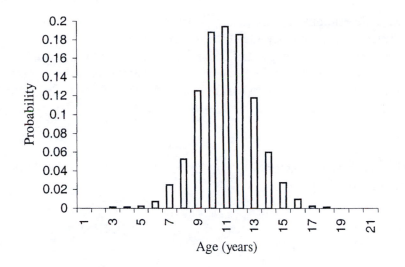

The data points are joined to generate a line graph that shows the probability distribution curve:

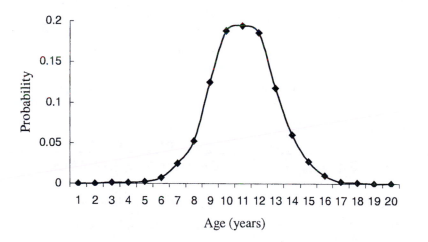

Note that if the number of records were very large, the probability distribution line graph would be a smoother curve or function, which would be the limiting probability distribution.

10.11 Mean, Median, and Mode of a Distribution

In addition to determining the frequency distribution, it is often necessary or desirable to characterize mathematically the distribution in terms of different parameters that describe its center, width, shape, and so on. The three commonly used parameters that measure the location of the center are *mean*, *median* and *mode*.

The mean, or average, of a distribution, as defined in Section 7.4, is the sum of all measurements divided by the number of measurements:

$$\bar{x} = \frac{\sum x}{N}.$$

The mean can also be determined by using the frequency distribution f or the probability distribution p, as follows:

$$\bar{x} = \frac{\sum fx}{N}$$

$$\bar{x} = \sum px.$$

Since each measurement is weighted by its frequency or probability, this technique of determining mean or average is also known as *weighted average*.

When the data of a measurement are sorted from lowest to highest, the *median* of the measurement represents the middle point of the data set. If there is an even number of data points, then it is the average of the two middle points. In situations where, because of many outlying measurements, the mean may be unreliable, the median is used.

The *mode* of a data set represents the measurement that occurs most often in the data set.

Note: *Mean* is the sum of the data divided by the number of data.

Median is the middle number when the data are arranged in numerical order.

Mode is the number(s) or item(s) that occurs most often. Remember the mnemonic: "*Mode occurs most often and median is in the middle.*"

Example 10.11.1
The ages (in years) of 15 students in a class are 17, 19, 17, 20, 22, 24, 28, 18, 19, 20, 24, 18, 18, 21, and 23. Determine the mean, mode, and median for the ages.

The total number of students, N, is 15. First, use the formula $\bar{x} = \dfrac{\Sigma x}{N}$ to determine the mean:

$$\text{Mean} = \frac{1}{15}\left[17+19+17+20+22+24+28+18+19+20+24+18+18+21+23\right] \text{ years}$$
$$= 20.5 \text{ years.}$$

You can also use the formula $\bar{x} = \Sigma px$ to determine the mean. To do so, find the frequency of occurrence (f) for each age and its corresponding probability ($p = f/N$). The results are as follows:

Age	17	18	19	20	21	22	23	24	25	26	27	28
f	2	3	2	2	1	1	1	2	0	0	0	1
p	0.13	0.20	0.13	0.13	0.07	0.07	0.07	0.13	0.00	0.0	0.0	0.07

Thus, the mean age is
$$\bar{x} = \Sigma px = 17 \times 0.13 + 18 \times 0.20 + 19 \times 0.13 + 20 \times 0.13 + 21 \times 0.07 + 22 \times 0.07 +$$

$$23 \times 0.07 + 24 \times 0.13 + 25 \times 0.0 + 26 \times 0.0 + 27 \times 0.0 + 28 \times 0.07 = 20.5 \text{ years.}$$

To determine the *mode*, look at the data to find which age or ages appear most often; that is, find the age for which the frequency f (and the probability p) is a maximum. It is clear that f and p are maximum for the age 18. Therefore, the mode is 18 years.

To determine the *median* age, first list the 15 ages in order of increasing numbers. Then find the middle number.

17   17   18   18   18   19   19   20   20   21   22   23   24   24   28

7 numbers   median   7 numbers

The middle number is 20. Therefore, the median age is 20 years.

Measures of the Shapes of a Distribution

The general shape of a curve or a distribution is characterized by the following two quantities:

- Skewness

- Kurtosis

Figure 10.10

Figure 10.11

The *skewness* of a distribution describes how much a curve is weighted to the left or to the right. This is explained in Figure 10.10. *Skew right* means that the tail of the curve is toward the right, and *skew left* means the tail of the curve is toward the left.

The *kurtosis* of a distribution is a measure of how sharp the peak of the distribution is. A distribution with high kurtosis has a pointy peak, while a distribution with low kurtosis has a flat peak (Figure 10.11).

Answers to Exercises in Chapter 10

1. Average value of $g = 9.79$ m/s^2
   0.2 % error
   a.d. = 0.04 m/s^2
   $\sigma = 0.05$ m/s^2
   A.D. = 0.01 m/s^2
   SDOM = 0.02 m/s^2

2. (a) (40.3 ± 0.4) cm
   (b) (307 ± 7) cm^2
   (c) (2846 ± 57) cm^2
   (d) 3.0 ± 0.1

6. (a) $\dfrac{1}{9}$

   (b) $\dfrac{1}{9}$

7. (a) 64
   (b) 6
   (c) 15
   (d) $\dfrac{1}{64}$

   (e) $\dfrac{3}{32}$

   (f) $\dfrac{15}{64}$

Appendix A

Important Conversion Factors

Length
1 mi =1.609 km
1 ft = 0.3048 m
1 m = 3.281 ft
1 in. = 0.0254 m
1 nautical mile = 1.852 km
1 fermi = 10^{-15} m
1 angstrom = 10^{-10} m
1 light-year = 9.46×10^{15} m
1 parsec = 3.09×10^{16} m

Mass
1 kg = 0.0685 slug
1 atomic mass unit (u) = 1.6605×10^{-27} kg

Time
1 day = 8.64×10^{4} s
1 year = 3.156×10^{7} s

Angle
1 radian (rad) = 57.3°
1° = 0.01745 rad

Area
1 in^2 = 6.45×10^{-4} m^2
1 ft^2 = 9.29×10^{-2} m^2

Volume
1 liter (L) = 1000 mL = 1000 cm^3 = 1.0×10^{-3} m^3
1 gallon = 4 qt = 3.78 L

1 m^3 = 35.31 ft^3

Speed
1 mi/h = 1.609 km/h = 0.447 m/s
1 ft/s = 0.305 m/s
1 km/h = 0.278 m/s
1 knot = 1.151 mi/h = 0.5144 m/s
1 rev/m = 0.1047 rad/s

Force
1 lb = 4.45 N
1 N = 0.225 lb
1 N = 10^5 dyne

Work and Energy
1 J = 10^7 ergs = 0.738 ft·lb
1 ft·lb = 1.36 J
1 kcal = 4.18×10^3 J = 3.97 Btu
1 kWh = 3.60×10^6 J
1 electron volt (eV) = 1.6022×10^{-19} J

Power
1 W = 0.738 ft·lb/s
1 hp = 550 ft·lb/s = 746 W

Pressure
1 atm = 1.013×10^5 N/m^2
1 lb/in^2 = 6.90×10^3 N/m^2
1 Pa = 1 N/m^2 = 1.45×10^{-4} lb/in^2

Temperature
$$F = \frac{9}{5} C + 32 \qquad F = \text{Fahrenheit}$$
$$C = \frac{5}{9} (F - 32) \qquad C = \text{Celsius}$$
$$K = C + 273.15 \qquad K = \text{Kelvins}$$

Appendix B

Fundamental Physical Constants

It has been found that certain numbers play important roles in describing the universe and its physical principles. These numbers are called *fundamental constants*. Following is a list of most fundamental constants, their symbols, and their values.

Constants	Symbol	Value
Speed of light	c	2.9979×10^8 m/s
Gravitational constant	G	6.6726×10^{-11} N·m²/kg²
Planck's constant	h	6.6261×10^{-34} J·s
$h/2\pi$	$\hbar$	1.05457×10^{-34} J·s
Boltzmann's constant	k	1.3807×10^{-23} J/K
Molar gas constant	R	8.3145 J/mol·K
Avogadro's number	N_A	6.0214×10^{23} mol⁻¹
Stefan-Boltzman constant	σ	5.6705×10^{-8} /m²·K⁴
Permittivity of free space	ε_0	8.8542×10^{-12} /N·m²
Permeability of free space	μ_0	1.2566×10^{-6} T·m/A
Charge of an electron	e	1.6022×10^{-19} C
Electron rest mass	m_e	9.1094×10^{-31} kg = .000549 u = 0.511 MeV/c^2
Proton rest mass	m_p	1.6726×10^{-27} kg = 1.00728 u = 938.3 MeV/c^2
Neutron rest mass	m_n	1.6750×10^{-27} kg = 1.008665 u = 939.6 MeV/c^2
Proton-electron mass ratio	m_p/m_e	1836.1527
Fine-structure constant, $\mu_0 c e^2/2h$	α	7.2974×10^{-3}
Inverse fine structure constant	α^{-1}	137.0360
Rydberg constant	R_∞	1.0974×10^{-7} m⁻¹
Faraday constant, $N_A e$	F	9.6485×10^{-4} C·mol⁻¹
Magnetic flux quantum, $h/2e$	Φ_0	2.06783×10^{-15} Wb

Appendix C

Useful Data and Values of Some Numbers

Useful Data

Absolute Zero (0 K) = –273.15 C

Earth: Mass = 5.97×10^{24} kg
                Radius (mean) = 6.38×10^{6} m
                Acceleration due to gravity (mean) on the surface = 9.8 m/s^2

Sun: Mass = 1.99×10^{30} kg
                Radius (mean) = 6.96×10^{8} m

Moon: Mass = 7.35×10^{22} kg
                Radius (mean) = 1.74×10^{6} m
                Acceleration due to gravity = 1.67 m/s^2

Earth-sun distance (mean) = 149.6×10^{9} m
Earth-moon distance (mean) = 384×10^{6} m

Values of Some Numbers

π = 3.1415927
e = 2.7182818

ln 2 = 0.6931472
$\log_{10}e$ = 0.4342945
ln 10 = 2.3025851

$\sqrt{2}$ = 1.4142136
$\sqrt{3}$ = 1.7320508

Appendix D

Graphical Method of Solving Two Equations

To solve two equations by the graphical method:

(i)   Graph each equation carefully.

(ii)   Determine the x- and y-coordinates of the point where the two graphs intersect. The coordinates of the point of intersection is the solution of x and y.

(iii)   Verify that x and y satisfy the original equations.

Graphical method is usually not used in introductory physics to solve equations. However, it is needed to solve two equations at intermediate level physics.

Example D1

Solve $\begin{cases} x - y = 2 \\ x + 2y = 8 \end{cases}$ by the graphical method.

For four different chosen values of x, different values of y are determined and shown below:

$x - y = 2$			$x + 2y = 8$	
x	y		x	y
0	−2		0	4
2	0		2	3
4	2		2	3
1	−1		−2	5

The x- and y-values of each equation are plotted and the points are joined to form two straight lines as shown in Figure D1.

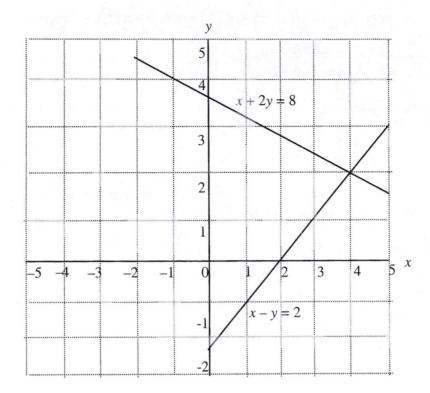

Figure D1 Cartesian graph of two equations.

The straight lines intersect at $x = 4$ and $y = 2$. Therefore, the solution is

$$\begin{cases} x = 4 \\ y = 2 \end{cases}$$

Exercises

1. Show graphically that the solutions of the equations $5x - 4y = 1$ and $3y = 5x - 2$ are $x = 1, y = 1$.

2. Show graphically that the solution of the transcendental equation $5e^{-x} + x - 5 = 0$ is $x \approx 5$. (*Hint*: Assume $f(x) = 5e^{-x}$ and $g(x) = 5 - x$, plot $f(x)$ and $g(x)$ determine where $f(x)$ and $g(x)$ intersect.)

*Appendix E

Solving Systems of Linear Equations: Method of Determinants (Cramer's Rule)

Cramer's rule provides a way for solving any set of n independent linear equations with n unknown variables. The procedure involves the calculation of the *determinants* of a *matrix*.

A matrix is a rectangular array of numbers or functions, such as $\begin{pmatrix} a_{11} & a_{12} & a_{13} & a_{14} \\ a_{21} & a_{22} & a_{23} & a_{24} \\ a_{31} & a_{32} & a_{33} & a_{34} \end{pmatrix}$.

The preceding matrix contains three rows (the number of horizontal lines in the array) and four columns (the number of vertical lines in the array). The *determinant* is defined only for square matrices, so that there are equal numbers of rows and columns. For example, for a matrix of order 2, $A = \begin{bmatrix} a_{11} & a_{12} \\ a_{21} & a_{22} \end{bmatrix}$, the determinant of the matrix is denoted as $|A|$ and is given by

$$|A| = \begin{vmatrix} a_{11} & a_{12} \\ a_{21} & a_{22} \end{vmatrix} = a_{11}a_{22} - a_{12}a_{21}.$$

In general, for a determinant of order n (that is, for a determinant containing n rows and n columns), the *minor* of an element a_{ij} of the determinant is defined as the determinant of order ($n-1$) that is obtained by deleting the ith row and the jth column (these are the row and column containing a_{ij}) from the original determinant. The minor of a_{ij} is denoted as $|M_{ij}|$. The *cofactor* of a_{ij}, usually denoted as Δ_{ij}, is defined as

$$\Delta_{ij} = (-1)^{i+j}|M_{ij}|.$$

For example, for the determinant $\begin{vmatrix} a_{11} & a_{12} & a_{13} \\ a_{21} & a_{22} & a_{23} \\ a_{31} & a_{32} & a_{33} \end{vmatrix}$,

$$\Delta_{11} = (-1)^{1+1}|M_{11}| = + \begin{vmatrix} a_{22} & a_{23} \\ a_{32} & a_{33} \end{vmatrix}.$$

302

The value of a determinant $|A|$ of order n is given by the sum of the n products obtained by multiplying each element in one row or column by its cofactor. As an example, expand the following determinant:

$$|A| = \begin{vmatrix} 2 & 6 & 2 \\ 3 & 8 & 1 \\ 4 & 5 & 6 \end{vmatrix} = (-1)^{1+1}\, 2 \begin{vmatrix} 8 & 1 \\ 5 & 6 \end{vmatrix} + (-1)^{1+2}\, 6 \begin{vmatrix} 3 & 1 \\ 4 & 6 \end{vmatrix} + (-1)^{1+3}\, 2 \begin{vmatrix} 3 & 8 \\ 4 & 5 \end{vmatrix}$$

$$= 2(48 - 5) - 6(18 - 4) + 2(15 - 32)$$

$$= 86 - 84 - 34 = -32.$$

Note that, although there are several possible ways of expanding a determinant, it can be shown that all these yield the same result. (Verify this.)

Now, consider any set of linear equations, for example, three linear equations with three unknowns:

$$a_{11}x + a_{12}y + a_{13}z = c_1$$

$$a_{21}x + a_{22}y + a_{23}z = c_2$$

$$a_{31}x + a_{32}y + a_{33}z = c_3.$$

The equations can be written in matrix form as

$$\begin{pmatrix} a_{11} & a_{12} & a_{13} \\ a_{21} & a_{22} & a_{23} \\ a_{31} & a_{32} & a_{33} \end{pmatrix} \begin{pmatrix} x \\ y \\ z \end{pmatrix} = \begin{pmatrix} c_1 \\ c_2 \\ c_3 \end{pmatrix}.$$

The solutions of x, y, and z, as given by Cramer's rule, are

$$x = \frac{\begin{vmatrix} c_1 & a_{12} & a_{13} \\ c_2 & a_{22} & a_{23} \\ c_3 & a_{32} & a_{33} \end{vmatrix}}{D} \qquad y = \frac{\begin{vmatrix} a_{11} & c_1 & a_{13} \\ a_{21} & c_2 & a_{23} \\ a_{31} & c_3 & a_{33} \end{vmatrix}}{D}$$

$$z = \dfrac{\begin{vmatrix} a_{11} & a_{12} & c_1 \\ a_{21} & a_{22} & c_2 \\ a_{31} & a_{32} & c_3 \end{vmatrix}}{D}, \qquad \text{where } D = \begin{vmatrix} a_{11} & a_{12} & a_{13} \\ a_{21} & a_{22} & a_{23} \\ a_{31} & a_{32} & a_{33} \end{vmatrix} \neq 0$$

The solution can be generalized for any number of equations and unknowns. If the denominator D is zero, the equations are not independent and Cramer's rule cannot be applied.

Cramer's rule is very useful in electricity to solve for electric currents in different sections of multiloop circuits.

Example (from electricity)

The electric currents (in milliampere, mA) i_1, i_2, and i_3 in different sections of a multiloop electric circuit are given by

$$i_1 + i_2 + i_3 = 0$$
$$2i_1 - i_2 + i_3 = -1$$
$$8i_1 - i_2 - 2i_3 = 11$$

Determine the currents.

Use Cramer's rule to solve for the currents. Comparing these equations with the preceding three linear equations, we find that the solutions are

$$i_1 = \dfrac{\begin{vmatrix} 0 & 1 & 1 \\ -1 & -1 & 1 \\ 11 & -1 & -2 \end{vmatrix}}{D}, \quad i_2 = \dfrac{\begin{vmatrix} 1 & 0 & 1 \\ 2 & -1 & 1 \\ 8 & 11 & -2 \end{vmatrix}}{D}, \text{ and, } i_3 = \dfrac{\begin{vmatrix} 1 & 1 & 0 \\ 2 & -1 & -1 \\ 8 & -1 & 11 \end{vmatrix}}{D}.$$

In this case,

$$D = \begin{vmatrix} 1 & 1 & 1 \\ 2 & -1 & 1 \\ 8 & -1 & -2 \end{vmatrix} = (-1)^{1+1}\,1 \begin{vmatrix} -1 & 1 \\ -1 & -2 \end{vmatrix} + (-1)^{1+2}\,1 \begin{vmatrix} 2 & 1 \\ 8 & -2 \end{vmatrix} + (-1)^{1+3}\,1 \begin{vmatrix} 2 & -1 \\ 8 & -1 \end{vmatrix}$$

$$= 1(2+1) - (-4-8) + (-2+8) = 3 + 12 + 6 = 21$$

$$\begin{vmatrix} 0 & 1 & 1 \\ -1 & -1 & 1 \\ 11 & -1 & -2 \end{vmatrix} = (-1)^{1+1} \cdot 0 \begin{vmatrix} -1 & 1 \\ -1 & -2 \end{vmatrix} + (-1)^{1+2} \cdot 1 \begin{vmatrix} -1 & 1 \\ 11 & -2 \end{vmatrix} + (-1)^{1+3} \cdot 1 \begin{vmatrix} -1 & -1 \\ 11 & -1 \end{vmatrix}$$

$$= 0 - (2 - 11) + (1 + 11) = 9 + 12 = 21$$

$$\begin{vmatrix} 1 & 0 & 1 \\ 2 & -1 & 1 \\ 8 & 11 & -2 \end{vmatrix} = (-1)^{1+1} \cdot 1 \begin{vmatrix} -1 & 1 \\ 11 & -2 \end{vmatrix} + (-1)^{1+2} \cdot 0 \begin{vmatrix} -1 & 1 \\ 11 & -2 \end{vmatrix} + (-1)^{1+3} \cdot 1 \begin{vmatrix} 2 & -1 \\ 8 & 11 \end{vmatrix}$$

$$= (2 - 11) + 0 - (22 + 8) = -9 + 30 = 21$$

$$\begin{vmatrix} 1 & 1 & 0 \\ 2 & -1 & -1 \\ 8 & -1 & 11 \end{vmatrix} = (-1)^{1+1} \cdot 1 \begin{vmatrix} -1 & -1 \\ -1 & 11 \end{vmatrix} + (-1)^{1+2} \cdot 1 \begin{vmatrix} 2 & -1 \\ 8 & 11 \end{vmatrix} + (-1)^{1+3} \cdot 0 \begin{vmatrix} 2 & -1 \\ 8 & -1 \end{vmatrix}$$

$$= (-11 - 1) - (22 + 8) + 0 = -12 - 30 = -42.$$

Therefore, the solutions are

$$i_1 = \frac{\begin{vmatrix} 0 & 1 & 1 \\ -1 & -1 & 1 \\ 11 & -1 & -2 \end{vmatrix}}{D} = \frac{21}{21} = 1 \text{ mA}, \quad i_2 = \frac{\begin{vmatrix} 1 & 0 & 1 \\ 2 & -1 & 1 \\ 8 & 11 & -2 \end{vmatrix}}{D} = \frac{21}{21} = 1 \text{ mA},$$

$$i_3 = \frac{\begin{vmatrix} 1 & 1 & 0 \\ 2 & -1 & -1 \\ 8 & -1 & 11 \end{vmatrix}}{D} = \frac{-42}{21} = -2 \text{ mA}.$$

*Appendix F

Components of a Vector in Different Systems in Terms of Cartesian Components

Polar system
$$A_r = \frac{x\,A_x + y\,A_y}{\sqrt{x^2 + y^2}}, \qquad A_\theta = \frac{-y\,A_x + x\,A_y}{\sqrt{x^2 + y^2}}$$

Spherical system
$$A_r = \frac{x\,A_x + y\,A_y + z\,A_z}{\sqrt{x^2 + y^2 + z^2}}, \qquad A_\theta = \frac{-y\,A_x + x\,A_y}{\sqrt{x^2 + y^2}},$$

$$A_\phi = \frac{xz\,A_x + yz\,A_y - \left(x^2 + y^2\right)A_z}{\sqrt{x^2 + y^2}\sqrt{x^2 + y^2 + z^2}}$$

Cylindrical system
$$A_r = \frac{x\,A_x + y\,A_y}{\sqrt{x^2 + y^2}}, \qquad A_\theta = \frac{-y\,A_x + x\,A_y}{\sqrt{x^2 + y^2}},$$

$$A_z = A_z$$

*Appendix G

Relationships between Unit Vectors in Different Coordinate Systems

Relationships between Unit Vectors in Cartesian and Polar Systems

$$\hat{i} = \cos\theta\,\hat{r} - \sin\theta\,\hat{\theta} \qquad\qquad \hat{r} = \cos\theta\,\hat{i} + \sin\theta\,\hat{j}$$

$$\hat{j} = \sin\theta\,\hat{r} + \cos\theta\,\hat{\theta} \qquad\qquad \hat{\theta} = -\sin\theta\,\hat{i} + \cos\theta\,\hat{j}$$

Relationships between Unit Vectors in Cartesian and Spherical Polar Systems

$$\hat{i} = \cos\theta\sin\phi\,\hat{r} - \sin\theta\,\hat{\theta} + \cos\theta\cos\phi\,\hat{\phi} \qquad \hat{r} = \cos\theta\sin\phi\,\hat{i} - \sin\theta\sin\phi\,\hat{j} + \cos\phi\,\hat{k}$$

$$\hat{j} = \sin\theta\sin\phi\,\hat{r} - \cos\theta\,\hat{\theta} + \cos\theta\sin\phi\,\hat{\phi} \qquad \hat{\theta} = -\sin\theta\,\hat{i} + \cos\theta\,\hat{j}$$

$$\hat{k} = \cos\theta\,\hat{r} - \sin\theta\,\hat{\phi} \qquad\qquad\qquad \hat{\phi} = \cos\theta\cos\phi\,\hat{i} + \cos\theta\sin\phi\,\hat{j} + \sin\phi\,\hat{k}$$

Relationships between Unit Vectors in Cartesian and Cylindrical Systems

$$\hat{i} = \cos\theta\,\hat{r} - \sin\theta\,\hat{\theta} \qquad\qquad \hat{r} = \cos\theta\,\hat{i} + \sin\theta\,\hat{j}$$

$$\hat{j} = \sin\theta\,\hat{r} + \cos\theta\,\hat{\theta} \qquad\qquad \hat{\theta} = -\sin\theta\,\hat{i} + \cos\theta\,\hat{j}$$

$$\hat{k} = \hat{k} \qquad\qquad\qquad\qquad \hat{k} = \hat{k}$$

*Appendix H

Rotation of the Coordinate Axes andTransformation of Vectors

Rotation of coordinate axes or transformation of vectors from one set of axes to another set of axes is not common in introductory physics. It is common in intermediate level physics such as in intermediate mechanics or relativity theory. Let (A_x, A_y) be the x- and y-components of a vector **A** with respect to the usual x- and y-axes. $(A_{x'}, A_{y'})$ are the components of the same vector **A** with respect to the new x'- and y'-axes. The coordinate axes x' and y' have the same origin as the x and y axes and the x'- and y'-axes have been rotated by an angle θ in the xy plane about their common z axis (not shown). The vector **A** makes angles α and β with the x- and x'-axes, respectively. The original and new axes and the vector **A** are shown in Figure H1. We want to know how does the vector transform when the coordinates axes are changed.

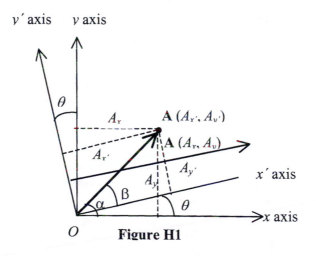

Figure H1

From figure H1 $A_x = A \cos \alpha$ and $A_y = A \sin \alpha$.

$A_x' = A \cos \beta = A \cos(\alpha - \theta) = A (\cos \alpha \cos \theta + \sin \alpha \sin \theta) = A_x \cos \theta + A_y \sin \theta$

$A_y' = A \sin \beta = A \sin(\alpha - \theta) = A (\sin \alpha \cos \theta - \cos \alpha \sin \theta) = -A_x \sin \theta + A_y \cos \theta$

The preceding two equations are also written in matrix notation as

$$\begin{pmatrix} A_{x'} \\ A_{y'} \end{pmatrix} = \begin{pmatrix} \cos\theta & \sin\theta \\ -\sin\theta & \cos\theta \end{pmatrix} \begin{pmatrix} A_x \\ A_y \end{pmatrix},$$

where $\begin{pmatrix} \cos\theta & \sin\theta \\ -\sin\theta & \cos\theta \end{pmatrix}$ is called the matrix of rotation. The first equation is obtained

by multiplying the first term of the first row of matrix with the first term of $\begin{pmatrix} A_x \\ A_y \end{pmatrix}$ and

by multiplying the second term of the first row of the matrix with the second term of

$\begin{pmatrix} A_x \\ A_y \end{pmatrix}$. Similarly, the second equation is obtained from the multiplication of the second

row of the matrix with $\begin{pmatrix} A_x \\ A_y \end{pmatrix}$.

Note that the length of the vector **A** must be the same regardless of the coordinate axes. That is, $A_x^2 + A_y^2 = A_{x'}^2 + A_{y'}^2$.

If the vector **A** represents a positione vector **r** wirh components (x, y) and (x', y') with respect to the unprimed and primed axes, respectively, then

$$\begin{pmatrix} x' \\ y' \end{pmatrix} = \begin{pmatrix} \cos\theta & \sin\theta \\ -\sin\theta & \cos\theta \end{pmatrix} \begin{pmatrix} x \\ y \end{pmatrix}.$$

In general, for rotation of a vector **A** in three dimensionss about an arbitrary axis,

$$\begin{pmatrix} A_{x'} \\ A_{y'} \\ A_{z'} \end{pmatrix} = \begin{pmatrix} R_{xx} & R_{xy} & R_{xz} \\ R_{yx} & R_{yy} & R_{yz} \\ R_{zx} & R_{zy} & R_{zz} \end{pmatrix} \begin{pmatrix} A_x \\ A_y \\ A_z \end{pmatrix}.$$

It is usually written in compact form as $A_{i'} = \sum_{j=1}^{3} R_{ij} A_{ij}$.

Exercise

1. Use the preceding expressionss for $A_{x'}$ and $A_{y'}$ to show that for the rotation of Cartesian axes $A_x^2 + A_y^2 = A_{x'}^2 + A_{y'}^2$.

*Appendix I
Line Element, Volume Element, Gradient, Divergence, Curl, and Laplacian in Spherical and Cylindrical Coordinate Systems

Spherical Coordinate System

$$\mathbf{dl} = dr\,\hat{\mathbf{r}} + rd\theta\,\hat{\boldsymbol{\theta}} + r\sin\theta\,d\phi\,\hat{\boldsymbol{\phi}} \qquad\qquad dV = r^2\sin\theta\,dr\,d\theta\,d\phi$$

$$\nabla f = \frac{\partial f}{\partial r}\hat{\mathbf{r}} + \frac{1}{r}\frac{\partial f}{\partial\theta}\hat{\boldsymbol{\theta}} + \frac{1}{r\sin\theta}\frac{\partial f}{\partial\phi}\hat{\boldsymbol{\phi}}$$

$$\nabla\cdot\mathbf{A} = \frac{1}{r^2}\frac{\partial}{\partial r}\left(r^2 A_r\right) + \frac{1}{r\sin\theta}\frac{\partial}{\partial\theta}\left(\sin\theta\,A_\theta\right) + \frac{1}{r\sin\theta}\frac{\partial A_\phi}{\partial\phi}$$

$$\nabla\times\mathbf{A} = \frac{1}{r\sin\theta}\left[\frac{\partial}{\partial\theta}\left(\sin\theta\,A_\varphi\right) - \frac{\partial A_\theta}{\partial\varphi}\right]\hat{\mathbf{r}} + \frac{1}{r}\left[\frac{1}{\sin\theta}\frac{\partial A_r}{\partial\phi} - \frac{\partial}{\partial r}\left(r A_\phi\right)\right]\hat{\boldsymbol{\theta}}$$
$$+\frac{1}{r}\left[\frac{\partial}{\partial r}\left(r A_\theta\right) - \frac{\partial A_r}{\partial\theta}\right]\hat{\boldsymbol{\phi}}$$

$$\nabla^2 f = \frac{1}{r^2}\frac{\partial}{\partial r}\left(r^2\frac{\partial f}{\partial r}\right) + \frac{1}{r^2\sin\theta}\frac{\partial}{\partial\theta}\left(\sin\theta\frac{\partial f}{\partial\theta}\right) + \frac{1}{r^2\sin^2\theta}\frac{\partial^2 f}{\partial\phi^2}.$$

Cylindrical Coordinate System

$$\mathbf{dl} = dr\,\hat{\mathbf{r}} + rd\theta\,\hat{\boldsymbol{\theta}} + dz\,\mathbf{k} \qquad\qquad dV = r\,dr\,d\theta\,dz$$

$$\nabla f = \frac{\partial f}{\partial r}\hat{\mathbf{r}} + \frac{1}{r}\frac{\partial f}{\partial\theta}\hat{\boldsymbol{\theta}} + \frac{\partial f}{\partial z}\mathbf{k}$$

$$\nabla\cdot\mathbf{A} = \frac{1}{r}\frac{\partial}{\partial r}\left(r A_r\right) + \frac{1}{r}\frac{\partial A_\theta}{\partial\theta} + \frac{\partial A_z}{\partial z}$$

$$\nabla\times\mathbf{A} = \left[\frac{1}{r}\frac{\partial A_z}{\partial\theta} - \frac{\partial A_\theta}{\partial z}\right]\hat{\mathbf{r}} + \frac{1}{r}\left[\frac{\partial A_r}{\partial z} - \frac{\partial A_z}{\partial r}\right]\hat{\boldsymbol{\theta}} + \frac{1}{r}\left[\frac{\partial}{\partial r}\left(r A_\theta\right) - \frac{\partial A_r}{\partial\theta}\right]\mathbf{z}$$

$$\nabla^2 f = \frac{1}{r}\frac{\partial}{\partial r}\left(r\frac{\partial f}{\partial r}\right) + \frac{1}{r^2}\frac{\partial^2 f}{\partial\theta^2} + \frac{\partial^2 f}{\partial z^2}$$

*Appendix J

Approximate Formulas for Definite Integrals

Let the interval from $x = a$ to $x = b$ be subdivided into n equal parts by the point $a = x_0, x_1, x_2, \cdots, x_{n-1}, x_n = b$. Let $y_0 = f(x_0), y_1 = f(x_1), y_2 = f(x_2), \cdots, y_n = f(x_n), h = (b - a)/n$ (Figure J1). Then any of the following formulas can be used to determine approximately the value of the definite integral $\int_0^a f(x)\,dx$.

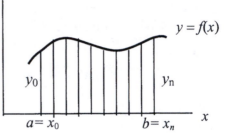

Rectangular Formula

$$\int_0^a f(x)\,dx \approx h\,(y_0 + y_1 + y_2 + \cdots + y_{n-1})$$

Trapezoidal Formula

$$\int_0^a f(x)\,dx \approx \frac{h}{2}\,(y_0 + 2y_1 + 2y_2 + \cdots + 2y_{n-1} + y_n)$$

Figure J1

Simpson's Formula for *n* Even

$$\int_0^a f(x)\,dx \approx \frac{h}{3}\,(y_0 + 4y_1 + 2y_2 + 4y_3 + \cdots + 2y_{n-2} + 4y_{n-1} + y_n)$$

*Appendix K
Common Probability Distributions

There are several important types of limiting distributions that are often found in physical problems or experiments. These include *Gaussian distribution*, *Poisson distribution*, *exponential distribution* and *Binomial distribution*. Table K1 lists the mathematical functions that characterize the probability distribution functions and the meaning of the terms in the expressions.

- *Gaussian distribution* is the most common distribution. It is characterized by a symmetric bell-shaped curve and is also called *normal distribution* or *error distribution*. If a large number of random, independent variables are measured in an experiment, the results form a Gaussian distribution symmetric with respect to the mean value of the result.

- *Binomial distribution* results in situations where there are only two possible outcomes, such as head or tail.

- *Random walk distribution* is a special situation of binomial distribution where the two possible outcomes have equal probabilities to occur (such as tossing a coin).

- *Poisson's distribution* occurs in the results of measurements of events that occur randomly but at a low, constant rate.

- *Exponential distribution* occurs in situations where waiting times are random but occurs at a constant average rate.

Table K1 Common Probability Distribution Functions in Physics

Gaussian Distribution

$$p(x) = \frac{1}{\sqrt{2\pi\sigma^2}} e^{-(x-\bar{x})^2/2\sigma^2}$$

σ = Width parameter
$\bar{x}$ = Center (or mean) of the distribution

Binomial Distribution

$$p(x) = \frac{N!}{x!(N-x)!} p^x q^{N-x}$$

p = Probability of the first outcome
$q = 1 - p$ = Probability of the second outcome
N = Number of events or trials
x = number of time the first outcome occurs

Random Walk Distribution

$$p(x) = \frac{1}{2^N} \frac{N!}{x!(N-x)!}$$

x = Number of time first outcome occurs
N = Number of events or trials

Poisson's Distribution

$$p(x) = \frac{\mu^N}{N!} e^{-\mu}$$

μ = Average rate of events

Exponential Distribution

$$p(x) = \lambda e^{-\lambda x}$$

λ = Average rate of waiting times

Example K1

When the error is random, the probability distribution of the result of a large number of measurements in an experiment is found to follow a Gaussian distribution $p(x) =$ $\left(1/\sqrt{2\pi\sigma^2}\right)e^{-(x-\bar{x})^2/2\sigma^2}$ with average $\bar{x} = 12.0$ units and width parameter $\sigma = 2.2$ units. Determine the probability that the result of a measurement, x, is (i) 11.0 units and (ii) 18.0 units

In this case, $\bar{x} = 12.0$ units and the width parameter $\sigma = 2.2$ units.

Therefore, $p(x) = \dfrac{1}{\sqrt{2\pi(2.2)^2}}e^{-(x-12.0)^2/2(2.2)^2}$.

When $x = 11.0$ units $\qquad\qquad p(10) = \dfrac{1}{\sqrt{2\pi(2.2)^2}}e^{-(11.0-12.0)^2/2(2.2)^2}$

Use a calculator to get the answer. $\qquad = 0.16$

When $x = 18.0$ units $\qquad\qquad p(18) = \dfrac{1}{\sqrt{2\pi(2.2)^2}}e^{-(18.0-12.0)^2/2(2.2)^2}$

Use a calculator to get the answer. $\qquad = 0.0044$

314

**Appendix L
Hyperbolic Functions

Hyperbolic functions are not as common as trigonometric functions in physics. However, they do appear in intermediate or advanced physics such as statistical physics, thermal physics, and solid-state physics. The hyperbolic functions have properties similar to trigonometric functions. The definitions of hyperbolic functions are straightforward.

Hyperbolic cosine of $x = \cosh x = \dfrac{e^x + e^{-x}}{2}$

Hyperbolic sine of $x = \sinh x = \dfrac{e^x - e^{-x}}{2}$

Like trigonometric cosine function, hyperbolic cosine (cosh) is an even function; that is, $\cosh(-x) = \cosh x$. Also, like trigonometric sine function, hyperbolic sine (sinh) is an odd function; that is $\sinh(-x) = -\sinh x$. An important property of sinh and cosh is that (compare with $\sin^2 x + \cos^2 x = 1$)

$$(\cosh x)^2 - (\sinh x)^2 = 1$$

It can be verified easily as follows:

$$(\cosh x)^2 - (\sinh x)^2 = \left(\frac{e^x + e^{-x}}{2}\right)^2 - \left(\frac{e^x - e^{-x}}{2}\right)^2$$

$$= \frac{1}{4}[(e^{2x} + 2\,e^x\,e^{-x} + e^{-2x}) - (e^{2x} - 2\,e^x\,e^{-x} + e^{-2x})]$$

$$= \frac{1}{4}(4\,e^x\,e^{-x}) = 1.$$

The other hyperbolic functions are defined in a similar way that the other trigonometric functions are defined.

$$\tanh x = \frac{\sinh x}{\cosh x} = \frac{e^x + e^{-x}}{e^x - e^{-x}} \qquad \operatorname{csch} x = \frac{1}{\sinh x} = \frac{2}{e^x - e^{-x}}$$

$$\coth x = \frac{\cosh x}{\sinh x} = \frac{e^x - e^{-x}}{e^x + e^{-x}} \qquad \operatorname{sech} x = \frac{1}{\cosh x} = \frac{2}{e^x + e^{-x}}$$

The plots of cosh, sinh and tanh and coth functions are shown in Figure L1.

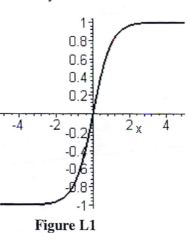

Figure L1

Note that the trigonometric functions $\sin t$ and $\cos t$ (t between 0 to 2π) parametrizes the unit circle $x^2 + y^2 = 1$, which can be verified easily by putting $x = \cos t$ and $y = \sin t$. The hyperbolic functions $\sinh t$ and $\cosh t$ (t between $-\infty$ to $+\infty$) parametrizes the hyperbola $x^2 - y^2 = 1$, which also can be verified easily by putting $x = \cosh t$ and $y = \sinh t$ (parametric equations are discussed in Chapter 4.)

Like trigonometric identities, there are similar identities for hyperbolic functions. For examples

$$\cosh (x \pm y) = \cosh x \cosh y \pm \sinh x \sinh y$$

$$\sinh (x \pm y) = \sinh x \cosh y \pm \cosh x \sinh y$$

List of other identities can be found in handbooks of mathematics.

Exercises

1. Start with right hand side and by direct substitution of the values of the cosh and sinh functions verify that the preceding relations of $\cosh(x \pm y)$ and $\sinh(x \pm y)$ are correct.

2. Show that $\tanh^2 x + \operatorname{sech}^2 x = 1$.

3. Show that :

   (a) $\dfrac{d}{dx} (\sinh x) = \cosh x$ (b) $\dfrac{d}{dx} (\cosh x) = \sinh x.$

4. Show that $A \cosh x + B \sinh x$ is a general solution of the differential equation $d^2 y/dx^2 - y = 0$.

Series Expansions for Hyperbolic Functions

   Hyperbolic functions like trigonometric functions can be expanded in power series containing infinite number of terms. The expansions are given below for sinh x, cosh x and tanh x:

$$\sinh x = x + \frac{x^3}{3!} + \frac{x^5}{5!} + \dots \qquad \text{for } |x| < \infty$$

$$\cosh x = 1 + \frac{x^2}{2!} + \frac{x^4}{4!} + \dots \qquad \text{for } |x| < \infty$$

$$\tanh x = x - \frac{x^3}{3} + \frac{2}{15} x^5 + \dots \qquad \text{for } |x| < \pi/2$$

Inverse Hyperbolic Functions

   Like inverse trigonometric functions, *inverse hyperbolic functions* are defined. If $x = \sinh y$, then $y = \sinh^{-1} x$. Similarly, other inverse hyperbolic functions are defined. For the properties of inverse hyperbolic functions see the CRC *Handbook of Tables for Mathematics* (CRC Press, New York) or any mathematics handbooks. The plots of inverse hyperbolic functions are shown in Figure L2.

317

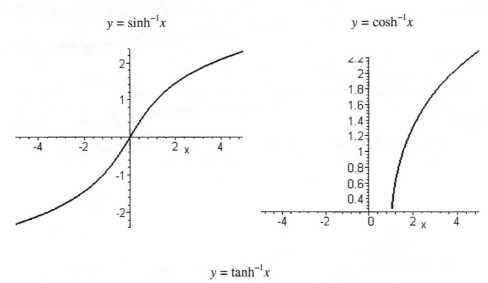

$y = \sinh^{-1}x$

$y = \cosh^{-1}x$

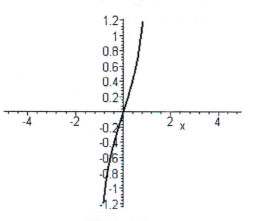

$y = \tanh^{-1}x$

Figure L2

318

Bibliography

1. R. David Gustafson and Peter D. Frisk, *Beginning and Intermediate Algebra 2*nd*: An Integrated Approach*. Pacific Grove, CA: Brooks/Cole, 1999, ISBN 0-534-35943-4.

2. Kristen A. Hubbard and Debora M. Katz, *The Physics ToolBox: A Survival Guide for Introductory Physics*. Pacific Grove, CA: Brooks/Cole, 2002, ISBN 0-03-034652-5.

3. Arnold D. Picker, *Preparing for General Physics*. Reading, MA: Addison-Wesley 1992, ISBN 0-201-56952-3.

4. James B. Seaborn, *Mathematics for the Physical Sciences*. New York: Springer, 2001, ISBN 0-387-95342-6.

5. David R. Lide (ed.), *CRC Handbook of Chemistry and Physics*. 77th edition. New York: CRC Press, 1996, ISBN 0-8493-0477-6.

6. Robert C. Weast (ed.), *CRC Handbook of Tables for Mathematics*, 4th edition. New York, CRC Press.

7. M. E. Shanks, C. F. Brumfiel, C. R. Fleenor and R. E. Eicholz, *Pre-Calculus Mathematics*. Reading, MA: Addison-Wesley, 1968.

8. Douglas C. Giancoli, *Physics for Scientists and Engineers*. 3rd ed. Upper Saddle River, NJ: Prentice Hall, 2000, ISBN 0-13-021517-1.

9. Eugene Hecht, *Physics: Calculus*. 2nd ed. Pacific Grove, CA: Brooks/Cole, 2000, ISBN 0-534-36270-2.

10. William Faissler, *An Introduction to Modern Electronics*. New York: John Wiley, 1991, ISBN 0-471-62242-7.

11. Raymond A. Serway, *Physics for Scientists and Engineers*. 3rd ed. Philadelphia, PA: Saunders College Publishing, 1992, ISBN 0-03-096027-4.

12. John D. Cutnell and Kenneth W. Johnson, *Physics*. 5th ed. New York: John Wiley, 2001, ISBN 0-471-32146-X.

13. James Walker, *Physics*. Upper Saddle River, NJ: Prentice Hall, 2002, ISBN 0-13-633124-6.

14. David J. Griffiths, *Introduction to Electrodynamics*, Englewood Cliffs, NJ: Prentice Hall, 1981.

15. Ronald K. Wangsness, *Electromagnetic Fields*, New York: John Wiley, 1979.

16. B. C. Das and B. N. Mukherjee, *Differential Calculus*, 43rd ed. Calcutta, India: U. N. Dhur and Sons.

17. K C Maity and P K Bagchi, *Integral Calculus including an Introduction to Diferential Equations*, 3rd ed. Calcutta, India: Books and Allied Private Limited.

18. M. R. Spiegel, *Schaum's Outline Series Theory and Problem of Mathematical Handbook of Formulas and Tables*. New York: McGraw Hill Book Company.